T0325668

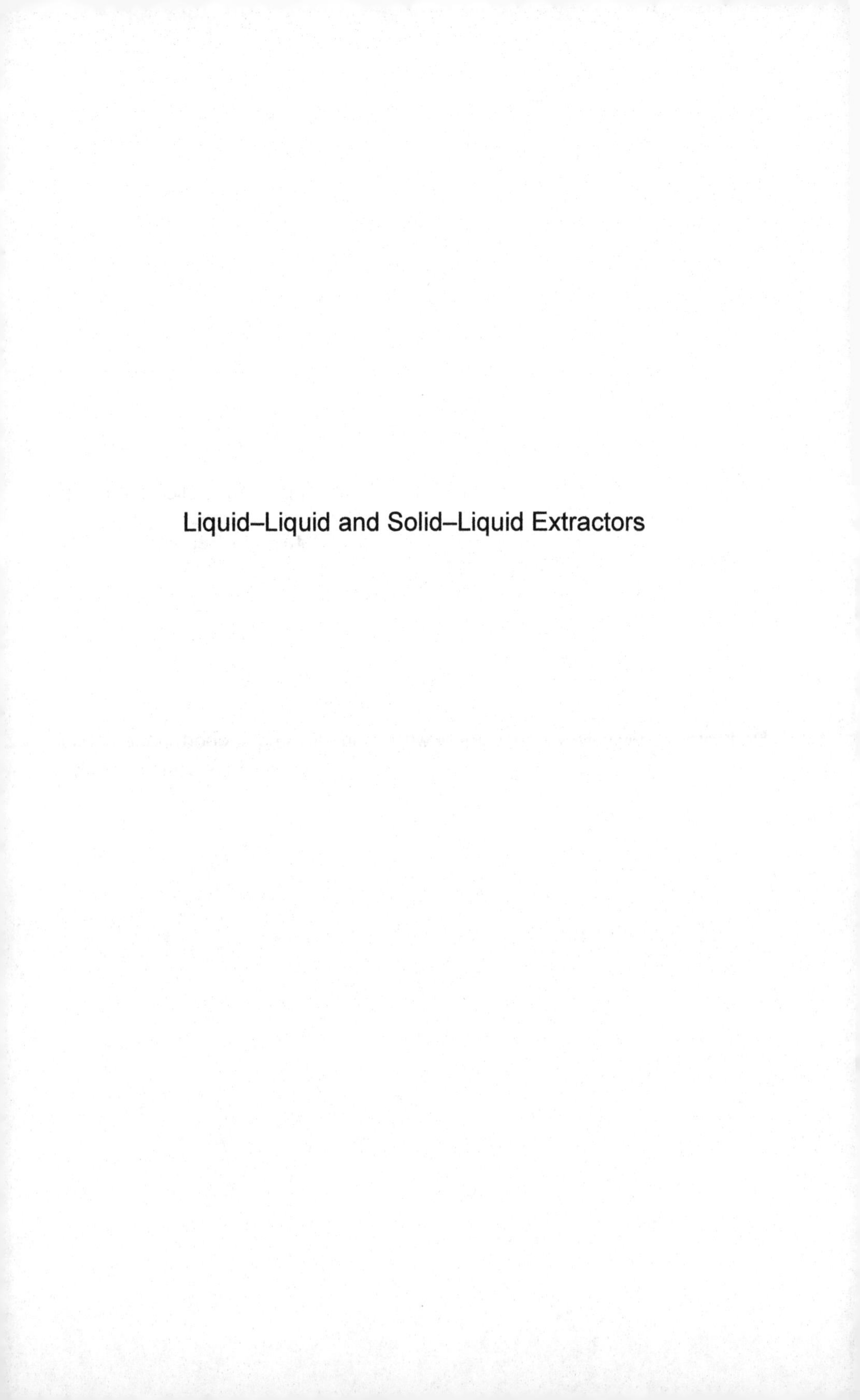

Liquid–Liquid and Solid–Liquid Extractors

There are no such things as applied sciences,
only applications of science.
Louis Pasteur (11 September 1871)

Dedicated to my wife, Anne, without whose unwavering support, none of this would have been possible.

Industrial Equipment for Chemical Engineering Set

coordinated by
Jean-Paul Duroudier

Liquid–Liquid and Solid–Liquid Extractors

Jean-Paul Duroudier

First published 2016 in Great Britain and the United States by ISTE Press Ltd and Elsevier Ltd

ISTE Press Ltd
27-37 St George's Road
London SW19 4EU
UK

www.iste.co.uk

Elsevier Ltd
The Boulevard, Langford Lane
Kidlington, Oxford, OX5 1GB
UK

www.elsevier.com

For information on all our publications visit our website at http://store.elsevier.com/

British Library Cataloguing-in-Publication Data
A CIP record for this book is available from the British Library
Library of Congress Cataloging in Publication Data
A catalog record for this book is available from the Library of Congress
ISBN 978-1-78548-178-9

Printed and bound in the UK and US

Contents

Preface

The observation is often made that, in creating a chemical installation, the time spent on the recipient where the reaction takes place (the reactor) accounts for no more than 5% of the total time spent on the project. This series of books deals with the remaining 95% (with the exception of oil-fired furnaces).

It is conceivable that humans will never understand all the truths of the world. What is certain, though, is that we can and indeed must understand what we and other humans have done and created, and, in particular, the tools we have designed.

Even two thousand years ago, the saying existed: "faber fit fabricando", which, loosely translated, means: "*c'est en forgeant que l'on devient forgeron*" (a popular French adage: *one becomes a smith by smithing*), or, still more freely translated into English, "practice makes perfect". The "artisan" (faber) of the 21st Century is really the engineer who devises or describes models of thought. It is precisely that which this series of books investigates, the author having long combined industrial practice and reflection about world research.

Scientific and technical research in the 20th Century was characterized by a veritable explosion of results. Undeniably, some of the techniques discussed herein date back a very long way (for instance, the mixture of water and ethanol has been being distilled for over a millennium). Today, though, computers are needed to simulate the operation of the atmospheric distillation column of an oil refinery. The laws used may be simple statistical correlations but, sometimes, simple reasoning is enough to account for a phenomenon.

Since our very beginnings on this planet, humans have had to deal with the four primordial "elements" as they were known in the ancient world: earth, water, air and fire (and a fifth: aether). Today, we speak of gases, liquids, minerals and vegetables, and finally energy.

The unit operation expressing the behavior of matter are described in thirteen volumes.

It would be pointless, as popular wisdom has it, to try to "reinvent the wheel" – i.e. go through prior results. Indeed, we well know that all human reflection is based on memory, and it has been said for centuries that every generation is standing on the shoulders of the previous one.

Therefore, exploiting numerous references taken from all over the world, this series of books describes the operation, the advantages, the drawbacks and, especially, the choices needing to be made for the various pieces of equipment used in tens of elementary operations in industry. It presents simple calculations but also sophisticated logics which will help businesses avoid lengthy and costly testing and trial-and-error.

Herein, readers will find the methods needed for the understanding the machinery, even if, sometimes, we must not shy away from complicated calculations. Fortunately, engineers are trained in computer science, and highly-accurate machines are available on the market, which enables the operator or designer to, themselves, build the programs they need. Indeed, we have to be careful in using commercial programs with obscure internal logic which are not necessarily well suited to the problem at hand.

The copies of all the publications used in this book were provided by the *Institut National d'Information Scientifique et Technique* at Vandœuvre-lès-Nancy.

The books published in France can be consulted at the *Bibliothèque Nationale de France*; those from elsewhere are available at the British Library in London.

In the in-chapter bibliographies, the name of the author is specified so as to give each researcher his/her due. By consulting these works, readers may gain more in-depth knowledge about each subject if he/she so desires. In a reflection of today's multilingual world, the references to which this series points are in German, French and English.

The problems of optimization of costs have not been touched upon. However, when armed with a good knowledge of the devices' operating parameters, there is no problem with using the method of steepest descent so as to minimize the sum of the investment and operating expenditure.

1

General Theory of Liquid–Liquid Extractors

1.1. Extraction by successive stages

1.1.1. *Feasibility of extraction (study using the ternary diagram)*

If we have the ternary diagram available to us, we can determine the quantity of solvent S needed to obtain a given extract and raffinate (linked by a tie line); see Figure 1.1.

Suppose we have taken the mass of the feed F and its composition (z_T and z_P, where $z_T + z_P = 1$).

In Figure 1.1, the lines RE and FS intersect at M, where the composition can be read from the diagram (the m_i values). We write that M is the barycenter (center of mass, which gives us the name "system of barycentric coordinates") of F and S:

$$M^* = F^* + S^*$$

so: $$(F + S)m = Fz + Sy$$

m, z, y, x: molar fractions of the chosen component in M^*, F^*, S^* and R^* or indeed:

$$S = F\frac{(m - z)}{(y - m)}$$

The calculation is subject to the following constraints:

– limiting quantities of solvent. The line SF intersects the isotherm of solubility at E_{max} and R_{min}, which correspond respectively to the maximum and minimum of the solvent. The only interesting region of the line is within the interval (E_{max}, R_{min}), in which the point M must fall. By successively placing the point M at E_{max} and R_{min}, it is easy to calculate S_{max} and S_{min};

– limiting composition of the feed. The line SF', which is a tangent to the solubility curve issuing from S, defines the point F' – i.e. the maximum concentration of solute T for there to be demixing.

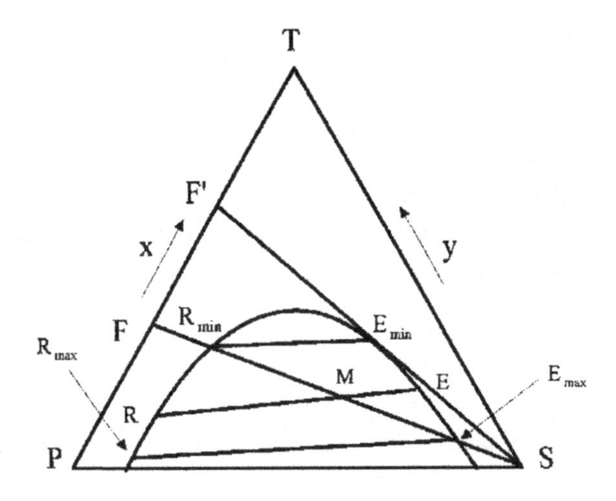

Figure 1.1. *Liquid–liquid equilibrium*

1.1.2. *Computer-based calculation of the solvent–raffinate equilibrium*

Refer to the discussion in section 2.3.

1.1.3. *Multi-stage countercurrent extraction*

Figure 1.2 shows the circulation of the fluids between the stages:

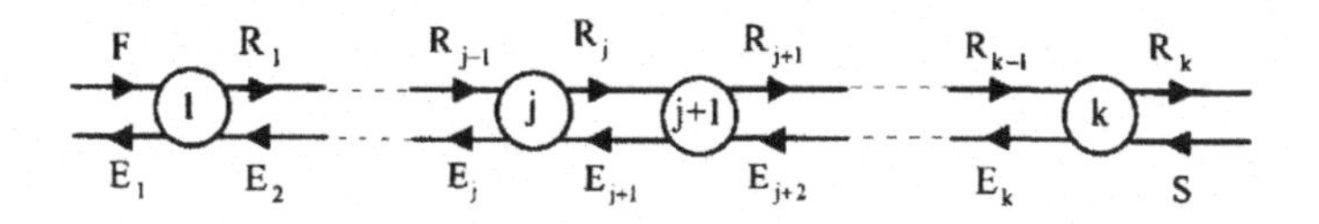

Figure 1.2. *Countercurrent stages*

The geometric construction which we shall now describe is indicated in Figure 1.3.

The overall balance of the device is written:

$$F^* + S^* = E_1^* + R_k^* = M^* \quad \text{(M: “additive” point)}$$

With this balance, if we know three of the four mixtures, we can determine the fourth. We can also write:

$$F^* - E_1^* = R_k^* - S^* = O^* \quad \text{(O: “subtractive” point)}$$

The point O is the pole of the construction.

Indeed, we can establish the balance:

$$F^* + E_{j+1}^* = E_1^* + R_j^*$$

so:

$$O^* = F^* - E_1^* = R_j^* - E_{j+1}^*$$

Thus, the points O, R_j and E_{j+1} are aligned.

Starting at E_1:

– the equilibrium gives R_1;

– the pole O gives E_2;

– the equilibrium gives R_2;

– the pole O gives E_3;

and so on.

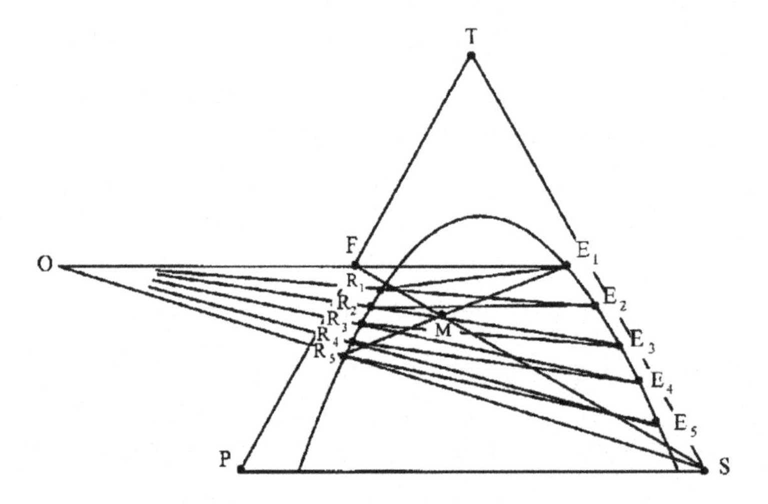

Figure 1.3. *Construction with the ternary diagram*

We can calculate the masses of extract and raffinate at each step.

$$R_{j-1} + E_{j+1} = R_j + E_j$$

Balance for the transferred solute:

$$R_{j-1}x_{j-1} + E_{j+1}y_{j+1} = R_j x_j + E_j y_j$$

By eliminating E_{j+1}, we find:

$$R_j = \frac{R_{j-1}(x_{j-1} - y_{j+1}) + E_j(y_{j+1} - y_j)}{x_j - y_{j+1}}$$

In addition:

$$E_{j+1} = R_j + E_1 - F$$

1.1.4. *Minimum quantity of solvent*

It is clear in Figure 1.3 that if a tie line passes through the pole 0, the above construction is unable to cross that line, and an infinite number of stages would be needed to reach it. This situation is known in distillation, and is called “pinching”.

Thus, we must avoid choosing R_k such that the line SR_k coincides with a tie line. Indeed, in this case, there would be pinching at the end of the installation where the solvent is introduced.

Remember that we have:

$$R_k^* = S^* + O^*$$

In other words, R_k is the barycenter of S and O. Thus, the further away point O is from R_k, the greater will be the quantity of solvent used.

From the above, we can deduce the method for finding the minimum quantity of solvent to be used: to do so, we extend the line segment R_kS on both sides. We also extend the tie lines $E_j R_j$ until their intersection O_j with the line R_kS. The point O which is furthest away from R_k corresponds to the minimum quantity of solvent to be used.

1.1.5. *Study using the distribution curve*

The balance of solute surrounding the feed-in end of stage j is written (see figure):

$$Rz + Ey_{j+1} = Rx_j + Ey_1$$

or indeed:

$$Rx_j + Ey_{j+1} = Rz - Ey_1 = Rx_k$$

This equation linking y_{j+1} to x_j defines the operating line. It is supposed that:

$$R = F = \text{const.} \qquad \text{and} \qquad E = S = \text{const.}$$

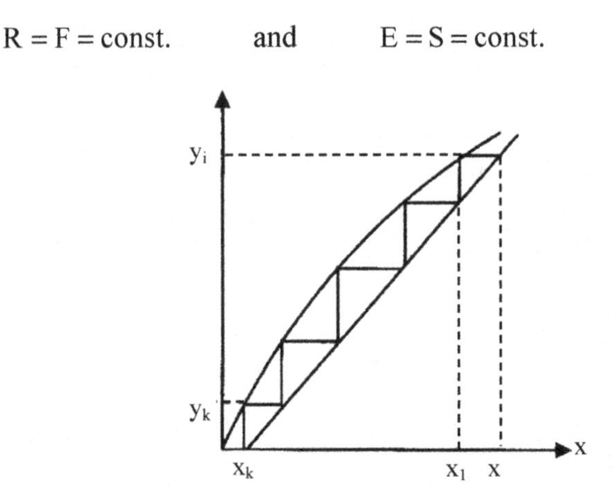

Figure 1.4. *Construction with the distribution curve*

1.1.6. *Aptitude of the system (solvent + product) for extraction*

Extraction is easier when:

1) for a molar fraction x in the residue of the product being transferred, the molar fraction y in the extract at equilibrium with x is high. Therefore, we need to maximize the sharing coefficient m, defined by:

$$m = \frac{dy}{dx}$$

2) for a given flowrate Q_R of raffinate, the flowrate Q_E of extract will be high. This is the same as maximizing:

$$Q_E/Q_R = U_E/U_R$$

U_E and U_R are the velocities in an empty bed of the extract and raffinate.

The “aptitude” of the system (solvent + product) combines these two parameters:

$$A = m \frac{U_E}{U_R}$$

In the theory of transfer unit heights, we could also use what we refer to as the "inaptitude" I of the system:

$$I = \frac{1}{A} = \frac{1}{m} \frac{U_R}{U_E}$$

1.2. Mixers-settlers

1.2.1. *The operation of mixing*

There have been numerous attempts to calculate, *a priori*, the effectiveness of a stage of mixing, but that effectiveness varies between 0.85 and 0.95, and only pilot testing on a 1/5 or 1/10 scale can help predict the behavior of the cascade.

Very often, the necessary lengths of stay in the mixer are less than one minute. Therefore, the cost of the installation will depend, above all, on the size of the decanters. Therefore, it is pointless to aim for violent agitation, which leads us to adopt:

– either the marine propeller;

– or the runner.

In a pilot test, we would examine, more specifically:

– the influence of the length of stay, i.e. of the flowrates, on the effectiveness and on the decantation;

– the homogeneity of the mixture, by taking samples from various places in the mixer.

For this purpose, we define a mixing index by the relation:

$$I_m = \frac{\phi}{Q_D/(Q_D + Q_C)}$$

ϕ : retention of dispersed phase;

Q_D and Q_C: volumetric flowrates of the dispersed phase and the continuous phase.

We try to ensure that the dispersion is perfect, meaning that I_m, which is always less than 1, comes close to this upper limit.

Note that the pilot must work on industrial products, because the effect of impurities is unpredictable.

1.2.2. *Separation of the two phases*

Depending on the products present, the coalescence of the dispersed phase can be achieved in two different ways:

– drop–drop coalescence,

– drop–interface coalescence.

In the former case, we need to use narrow, deep decanters.

In the latter case, we choose a flat decanter with a large horizontal surface area.

In solvent-based extraction, it is the second case which arises.

When a drop approaches the interface, a thin film of continuous phase separates it from the interface.

This film becomes thinner and eventually ruptures. The process is quicker when the viscosity of the continuous phase is lower.

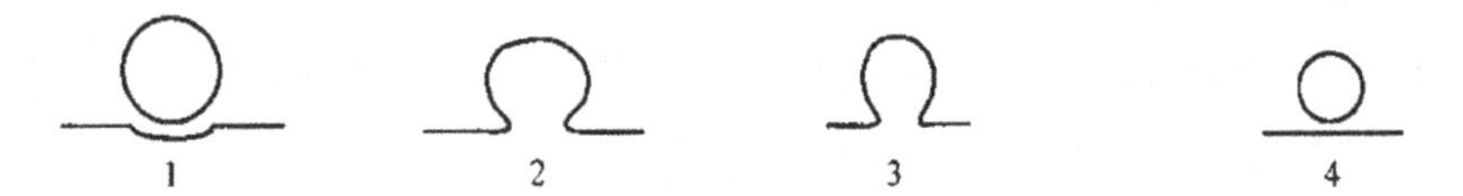

Figure 1.5. *Drop–interface coalescence*

We then obtain forms 2 and 3. The height of the cylinder remains constant, although its diameter decreases. Beyond a certain limit, the cylinder becomes unstable and shrinks to a point of its height, which releases a small secondary drop marked 4 in Figure 1.5.

In the case that the interfacial tension is low and the primary drop is large, it is not one but several secondary drops which are ejected.

If the primary drop embeds itself in the interface by less than 0.2 times its own diameter d, it coalesces by simple drainage.

If the embedding surpasses 0.26 d, we see the ejection of the secondary drop.

1.2.3. *Calculation for settlers*

An *a priori* calculation of the surface of a decanter would be illusory, which is why pilot tests are necessary.

Considerations similar to those developed for gravity-based liquid–solid settlers yield a simple relation between the surface Σ of the settler and the flowrate of the feed Q_A.

$$\Sigma = Q_A / V \quad \text{(V: decantation rate)}$$

The variable is the settling rate and, to enable the two phases to calm before they are withdrawn, it is taken as being equal to half the rate measured in a test tube.

In general, the length of stay in a settler is between 1 and 3 minutes. In a mixer-settler, 70% of the floor surface is occupied by the decanter and 30% by the mixer. The length of stay in the mixer is around a minute.

Practically, the principle of a settler is illustrated in Figure 1.6. In general, we find the level Z_1 of withdrawal of the light fraction by overflow and impose the level Z_3 of the interface. From this, we can deduce the overflow level Z_2 of the heavy fraction. If ρ_L and ρ_ℓ are the densities of the heavy fraction and the light fraction, let us write the expression of the driving pressure $\rho_L Z_2 g$.

$$\rho_L Z_2 = (Z_1 - Z_3)\rho_l \quad + \quad \frac{\Delta P_l}{g} \quad + \quad Z_3\rho_L \quad + \quad \frac{\Delta P_L}{g}$$

Height of light fraction	ΔP of light fraction in the pipe	Height of heavy fraction	ΔP of heavy fraction in the pipe

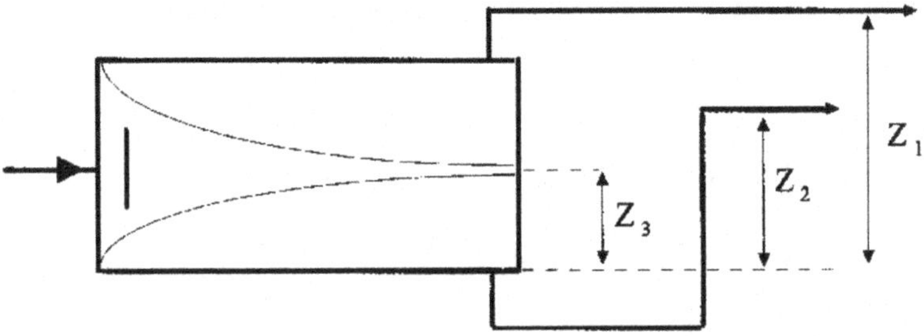

Figure 1.6. *Levels in the decanter*

The emulsion is input against a deflector so as not to disturb the calm which must be preserved in the settler.

Of course, whilst the static decanter is very widely used, it is still possible, sometimes, to use a hydrocyclone or even a centrifuge.

The phenomenon of settling plays a part in the choice of the contiguous phase by way of:

– the settling rate;

– the tendency to form a stable emulsion;

– the entrainment.

Entrainment generally takes place by imprisonment, within the droplets, or small sacs of the contiguous phase.

We may also see the formation of fouling at the interfaces of decanters. These phenomena are significant for the terminal stages of cascades.

1.3. Equation of concentrations for differential extractors

1.3.1. *Principle of differential extractors*

Instead of taking place in a series of distinct stages, differential extraction takes place in a vertical device where the surface of contact between the raffinate and the extract is distributed across the whole height of the device.

One of the two liquid phases is dispersed as drops which, themselves, are bathed in the contiguous phase. Later on, we shall see the criteria that govern the choice of a liquid for a given phase.

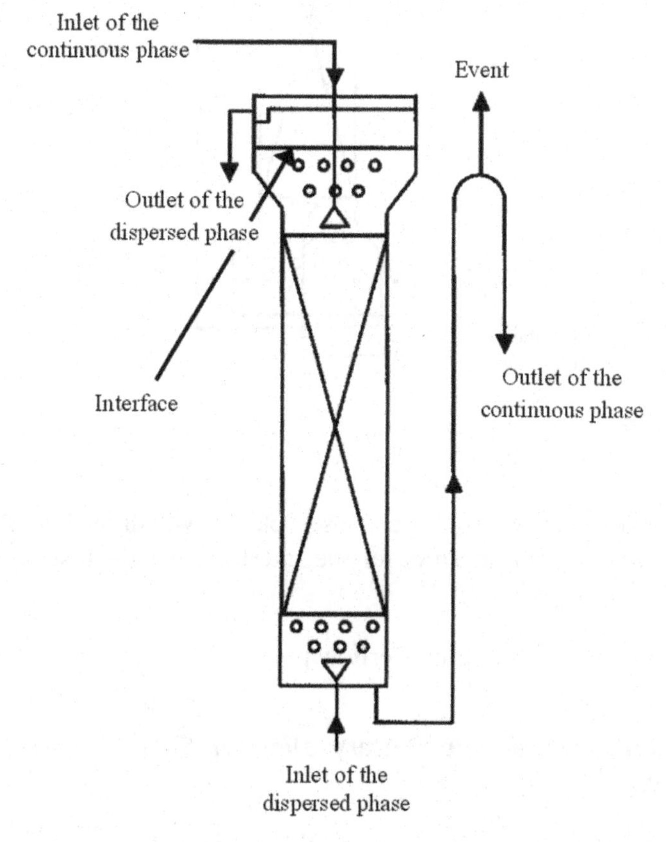

Figure 1.7. *Lightweight dispersed phase (the drops rise)*

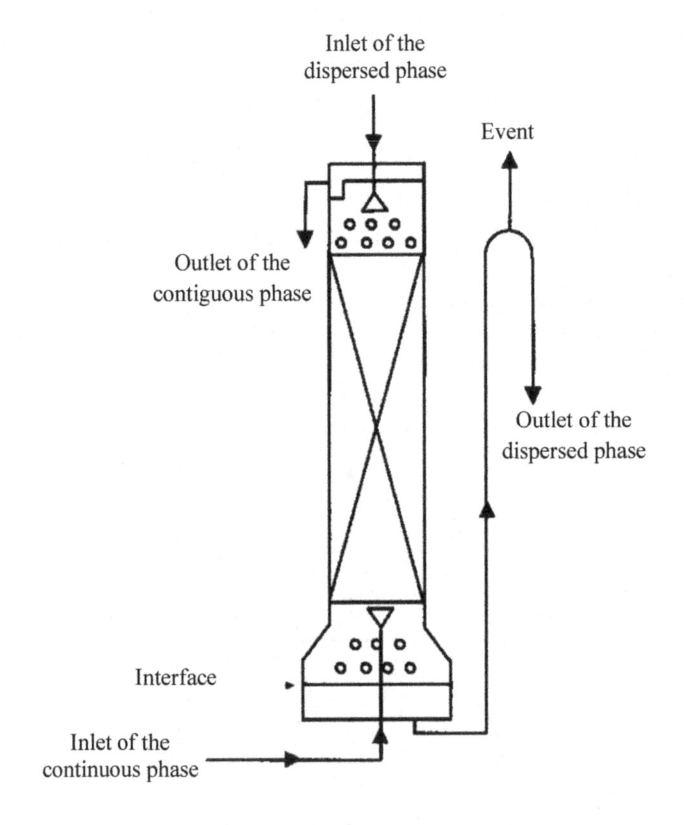

Figure 1.8. *Dispersed heavy phase (the drops descend)*

By adjusting the level of the gooseneck for withdrawal of the heavy phase, we can vary the position of the interface at will, and set it to the correct position.

The light phase flows by an overflow pipe.

1.3.2. *Balance of an elementary slice of the extractor with no backmixing*

The slice with index k is between the sections $k-1$ and k. The flux densities N_i are orientated positively for a transfer from the phase R (raffinate or residue) to the phase E (extract). The index k grows in the direction of progression of the extract.

The molar quantity of component i transferred in the slice k is:

$$w_{k,i} = N_{k,i} a_e \Delta V$$

$N_{k,i}$: flux density: $kmol.m^{-2}.s^{-1}$;

a_e: volumetric area for the transfer: m^{-1};

ΔV: volume of the slice k: m^3.

The total transferred quantity is:

$$W_k = \sum_i w_{k,i} \quad W_k > 0$$

Thus, we have the following molar flowrates on exit from the slice k:

$$E_k = E_{k-1} + W_k \quad \text{with} \quad y_{k,i} = \frac{E_{k,i}}{E_k}$$

$$R_k = R_{k-1} + W_k \quad \text{with} \quad x_{k,i} = \frac{R_{k,i}}{R_k}$$

The + sign in this last equation is justified because the index *k increases in the opposite direction to the progression of the liquid R.* The index k increases as we move from the poorer end to the richer end in terms of solute.

NOTE.–

We can write:

$$Q_R dc_{k,i,R} = w_{k,i} = N_{k,i} a_e A_C dz$$

The c values are the concentrations ($kmol.m^{-3}$).

Q_R and Q_E are the volume flowrates ($m^3.s^{-1}$).

Let us set:

$$U_R = Q_R / A_C \qquad \text{(and } U_E = Q_E / A_C \text{)}$$

Thus:

$$\frac{dc_{k,i,R}}{dz} = \frac{N_{k,i} a_e}{U_R} \quad \text{and, similarly,} \quad \frac{dc_{k,i,E}}{dz} = \frac{N_{k,i} a_e}{U_E}$$

1.3.3. *Axial dispersion (backmixing)*

Axial dispersion, which is commonly known as backmixing, tends to homogenize the two phases along the length of the device. In some cases, the device may be equivalent to a single extraction stage. In practice, backmixing is a crucial factor to take into consideration in the simulation of a differential extractor.

In general, backmixing is more intense in the continuous phase, because the drops carry a certain volume of that phase along with them in their wake and, more generally, the continuous phase is disturbed by the dispersed phase.

Axial dispersion is characterized by a parameter D_A, which has the dimensions of a diffusivity ($m^2.s^{-1}$), and is known as the "diffusivity of backmixing" or, more often, "axial dispersion coefficient" – i.e. dispersivity.

To account for backmixing, numerous authors propose to use a Péclet number that is invariable along the extractor. For the phase X:

$$Pé = \frac{LU_x}{D_x}$$

In this expression, the length L is a characteristic dimension of the device used, and is therefore constant along the length of the extractor. It is therefore acceptable to use a constant value for the ratio U_x / D_x. The problem, then, is one of choosing the right value for the velocity in an empty bed U_x. We propose to take, for U_x, an arithmetic mean between the extremities of the extractor.

$$\bar{U}_x = \frac{1}{2}\left(U_{x\ \mathrm{input}} + U_{x\ \mathrm{output}}\right)$$

The dispersivity D_x results from correlations which we shall use in the three examples of calculations for extractors, given in Chapter 2.

Backmixing hinders the transfer by decreasing the molar concentration $c_{R,i}$ in the phase R and increasing the molar concentration $c_{E,i}$ in the phase E. This is equivalent to decreasing the gap $(\mu_{R,i} - \mu_{E,i})$ between the chemical potentials:

$$c_{R,i} = c^*_{R,i} - \frac{D_R}{U_R}\frac{dc_{R,i}}{dz} \quad \text{and} \quad c_{E,i} = c^*_{E,i} + \frac{D_E}{U_E}\frac{dc_{E,i}}{dz} \qquad [1.1]$$

The differentials are positive because we are moving (increasing z) *from the poorer end toward the richer end*. More specifically, we know that:

$$\frac{dc^*_{R,i}}{dz} = \frac{a_e N_i}{U_R} \qquad \frac{dc^*_{E,i}}{dz} = \frac{a_e N_i}{U_E} \qquad [1.2]$$

The asterisk characterizes the transfer without backmixing.

To eliminate $c^*_{R,i}$ and $c^*_{E,i}$, let us differentiate equations [1.1] with respect to z and use equations [1.2]. We find:

$$\frac{dc_{R,i}}{dz} = \frac{a_e N_i}{U_R} - \frac{D_R}{U_R}\frac{d^2 c_{R,i}}{dz^2} \quad \text{and} \quad \frac{dc_{E,i}}{dz} = \frac{a_e N_i}{U_E} + \frac{D_E}{U_E}\frac{d^2 c_{E,i}}{dz^2}$$

In these equations, the velocities in an empty bed U_R and U_E are, by definition, positive, and so too are the dispersivities D_R and D_E.

These equations are of the form:

$$\alpha c'' + c' = \gamma \qquad \text{(concentration equation)}$$

$$\alpha_R = \frac{D_R}{U_R} > 0 \qquad \alpha_E = -\frac{D_E}{U_E} < 0$$

$$\gamma_R = \frac{a_e N_i}{U_R} > 0 \qquad \gamma_E = \frac{a_e N_i}{U_E} > 0$$

γ_E and γ_R are of the same sign and positive.

Note that if we were to travel along the column in the direction of decreasing concentration, we would necessarily have:

$$\gamma < 0 \qquad \alpha_R < 0 \qquad \alpha_E > 0 \qquad N_i < 0$$

As is the case for the calculation without backmixing, we can divide the extractor into elementary slices. The slice k lies between sections $k-1$ and k.

Let us set:

$$c' = \frac{dc}{dz} \quad \text{and} \quad c'' = \frac{d^2c}{dz^2}$$

The solution to the concentration equation on crossing a slice is:

$$c_{i,k} = c_{i,k-1} + \gamma\,(z_k - z_{k-1}) - \alpha\,(\gamma - c'_{i,k-1})\left[1 - \exp\left[-\frac{(z_k - z_{k-1})}{\alpha}\right]\right]$$

$$c'_{i,k} = \gamma - (\gamma - c'_{i,k-1})\,\exp\left[-\frac{(z_k - z_{k-1})}{\alpha}\right]$$

$$c''_{i,k} = \frac{1}{\alpha}\,(\gamma - c'_{i,k-1})\,\exp\left[-\frac{(z_k - z_{k-1})}{\alpha}\right]$$

We can check that, for $z_k = z_{k-1}$, we do indeed have:

$$c_{i,k} = c_{i,k-1} \quad \text{and} \quad c'_{i,k} = c'_{i,k-1}$$

These equations are satisfied even if the values of α and γ are different for slices $k-1$ and k.

The values of γ vary throughout the length of the extractor, and we have to write $\gamma_{i,k}$.

Also, using the expression of $c'_{i,k}$:

$$(\gamma_{i,k} - c'_{i,k}) = (\gamma_{i,k-1} - c'_{i,k-1}) \exp\left[-\left(\frac{z_{i,k} - z_{i,k-1}}{\alpha_k}\right)\right]$$

Thus, throughout a calculation, the sign of $(\gamma - c')$ remains constant. The same cannot be said of the sign of c''. Indeed, in view of the concentration equation, we know that:

$$c'' = \frac{\gamma - c'}{\alpha}$$

Upon exiting the extractor, the two fluids, as we shall see, have a zero value of c'. Therefore, the sign of $(\gamma - c')$ is identical to that of γ.

Figures 1.9 and 1.10 illustrate these results, with the simplistic hypothesis that the γ values are constant everywhere in the extractor.

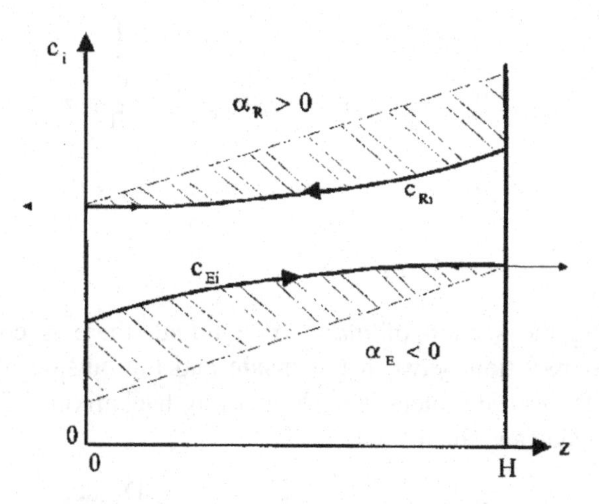

Figure 1.9. *Concentration profiles for* $\gamma > 0$

In each liquid, we solve as many equations as there are components, and accept that the coefficient of axial dispersion has a *common value* for all the components of the liquid.

The above supposes that the concentrations of the transferred components are low. If this is not so, we can take the following approach:

– after calculating the transfers in the slice in question with the above method, we take account of the presence of non-transferable inerts by calculating the new total molar flowrate and the new molar fractions on exiting the slice by keeping the values of $c' = \partial c / \partial z$ or $x' = \partial x / \partial z$;

– based on the values thus obtained, we run the calculations for the next slice.

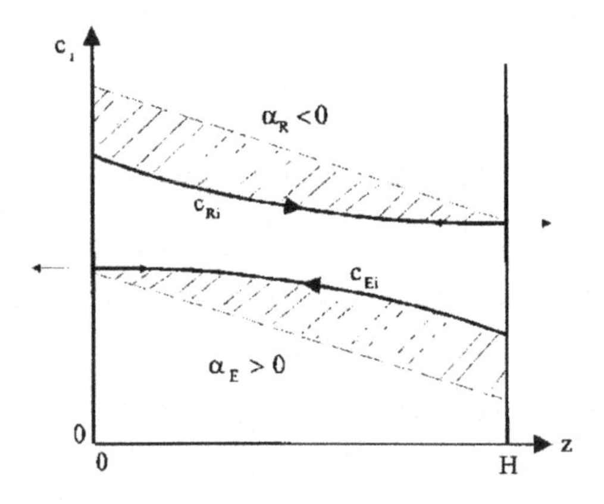

Figure 1.10. *Concentration profiles for* $\gamma < 0$

1.3.4. *Boundary conditions*

On crossing the section of *outlet* of a liquid, there is continuity and equality of composition between the inside and the outside of the device. However, on the outside, there is no longer any backmixing. Consequently, both on the outside and the inside:

Outlet of the phase R: $\quad c_{Ri} = c_{Ri}^* \quad$ so $\quad \dfrac{D_R}{U_R} \dfrac{dc_{Ri}}{dz} = 0$

Outlet of the extract E: $\quad c_{Ei} = c_{Ei}^* \quad$ so $\quad \dfrac{D_E}{U_E} \dfrac{dc_{Ei}}{dz} = 0$

On the other hand, at the *input* of the two liquids, the composition of each feed differs from that of the liquid present in the extractor. There is a "jump" in concentration.

1.3.5. *Overall balance of a component*

Let us assign the index 0 to the poor end (input) of the solvent and the index 1 to the rich end. For the component i:

$$\underbrace{(Uc)_{E,i,F} + (Uc)_{R,i,F}}_{\text{inlet}} = \underbrace{(Uc)_{E,i,1} + (Uc)_{R,i,0}}_{\text{outlet}}$$

The inlets have the index F instead of 0 or 1, because there is a concentration jump between the incoming fluid (fed in) and the fluid of the same nature but inside the extractor.

In that balance, the inlets are known, as is $(Uc)_{R,i,0}$, which is the term defining the specification of the component i – that is, its *maximum* concentration in the outgoing liquid, whether it is impurity in a "*raffinate*" or a noble product lost in a "*residue*".

Thus, this balance enables us to calculate $(Uc)_{E,i,1}$.

1.3.6. *Calculation procedure*

For the slice k, the liquid A and the component i, let us define the mean values:

$$\overline{x_{A,i,k}} = \frac{1}{2}(x_{A,i,k-1} + x_{A,i,k}) \qquad \text{with} \qquad x_i = \frac{c_i}{\sum_{j=1}^{n} c_j}$$

Based on these mean values, we calculate the chemical potentials and, consequently, the flux densities $N_{i,k}$. Let us take:

$$\gamma_{A,i,k} = a_e N_{i,k} / U_{A,k}$$

The calculation of the section k on the basis of the section $k-1$ takes place in two or three iterations because, to begin with, we do not know the $x_{A,i,k}$, and therefore have to take:

$$\overline{x_{A,i,k}}^{(0)} = x_{A,i,k-1}$$

The result is the following sequence of calculations.

1) A calculation without backmixing gives us the two concentration profiles $P_E^{(0)}$ and $P_R^{(0)}$.

2) We start at the outlet of the phase R, adopting the hypothesis that backmixing occurs only in that phase. We obtain the profile $P_R^{(1)}$.

3) We start at the outlet of the extract, working on the assumption that there is no backmixing in the phase R, which gives us the concentration profile $P_E^{(1)}$.

4) We start at the outlet of the phase R, using the profile $\frac{1}{2}(P_E^{(0)} + P_E^{(1)})$, which gives us the profile $P_R^{(2)}$.

5) We start at the outlet of the phase E, using the profile $\frac{1}{2}(P_R^{(0)} + P_R^{(1)})$, which gives us the profile $P_E^{(2)}$.

2n) We start at the outlet of the phase R, using the profile $\frac{1}{2}(P_E^{(n-2)} + P_E^{(n-1)})$, which gives us the profile $P_R^{(n)}$.

$2n+1$) We start at the outlet of the phase E, using the profile $\frac{1}{2}(P_R^{(n-2)} + P_R^{(n-1)})$, which gives us the profile $P_E^{(n)}$.

etc.

During the course of the calculation, it may be that the difference in chemical potentials $(\mu_{R,i,k} - \mu_{E,i,k})$ tends toward zero. We then say that there is "pinching". The remedy to this situation is to increase the flowrate of solvent.

NOTE.–

We can also operate as follows: to begin with, a material balance is found, which gives us the two output concentrations as a function of the inlet conditions. The calculation is done starting at the poor end (see Figure 1.10) and having decreased the concentration C_E of the incoming solvent by 15%. We reach the rich end when the derivative dc_E/dz is close to zero material. We then compare the outlet concentration thus found which is the outlet concentration given by the material balance. The relative gap between the two values can be used to correct the rate of 15% accepted for the concentration of the solvent at the inlet of the column. In principle, two or three iterations should be enough.

1.4. Transfer parameters

1.4.1. *Mean velocities in an empty bed*

The variations in molar flowrate in the slice k are:

$$\Delta W_{E,k} \quad \text{and} \quad \Delta W_{R,k}$$

The total concentrations $c_{E,T}$ and $c_{R,T}$ can generally be calculated by $c_T = 1/\sum_i x_i \overline{v}_i$, where $\overline{v}_i$ are the partial molar volumes that can be assimilated to the volumes in the pure state, except in cases such as water–alcohol mixtures.

The mean values of the velocities in an empty bed would therefore be:

$$\overline{U}_{E,k} = U_{E,k-1} + \frac{1}{2}\frac{W_{E,k}}{c_{E,T}A_c} \quad \text{and} \quad \overline{U}_{R,k} = U_{R,k-1} + \frac{1}{2}\frac{W_{R,k}}{c_{R,T}A_c}$$

1.4.2. *Mean drop diameter*

The aim is to find a way to simply express the surface of the drops for a unit of volume (or mass) of the dispersed phase.

In the domain of liquid–liquid extraction, the technique is to take photographs of the dispersion and count the drops. Let n_i represent the number of drops whose diameter is equal to $d_i \pm \delta d_i$. Let n_T be the total number of drops counted.

$$n_T = \sum_i n_i$$

The total surface of these drops is:

$$\pi \sum_i n_i d_i^2$$

Their volume is:

$$\frac{\pi}{6} \sum_i n_i d_i^3$$

By definition of d_{32}, the ratio sought is:

$$\sigma = \frac{6 \sum_i n_i d_i^2}{\sum_i n_i d_i^3} = \frac{6}{d_{32}}$$

Thus:

$$d_{32} = \frac{\sum_i n_i d_i^3}{\sum_i n_i d_i^2}$$

1.4.3. *Equivalence of d_{32} and the mean harmonic diameter*

In the operations of filtration, crystallization, adsorption, grinding, etc., we examine the particle-size distribution of the divided solid, which we process using the superposed sieves technique. The aperture (mesh dimension) of the sieves decreases as we move down the column. In the top sieve, we deposit a sample of the population being studied, and vibrate the whole setup. Finally, we weigh the "refuse" left in each sieve.

We shall suppose that all the particles have the same density and that, consequently, the mass and volume fractions are equal to a common value m_i.

The volume fraction of the class i in the total population is:

$$m_i = \frac{n_i \, \pi d_i^3/6}{\sum_j n_j \, \pi d_j^3/6} = \frac{n_i d_i^3}{\sum_j n_j d_j^3}$$

The mean harmonic diameter is defined by:

$$\frac{1}{d_h} = \sum_i \frac{m_i}{d_i} = \frac{1}{\sum_j n_j d_j^3} \sum_i n_i d_i^3 \times \frac{d_i^2}{d_i^3}$$

Thus:

$$\frac{1}{d_h} = \frac{\sum_i n_i d_i^2}{\sum_j n_j d_j^3} = \frac{1}{d_{32}}$$

1.4.4. *Interfacial area*

The surface of the drops for a unit volume of the dispersed phase is:

$$\sigma = \frac{6}{d_{32}}$$

The same surface for the unit volume of the mixture (dispersed phase + contiguous phase) is:

$$a = \frac{6\phi}{d_{32}}$$

ϕ: retention of the dispersed phase (i.e. volume of the dispersed phase divided by the volume of the dispersion).

1.4.5. *Transfer coefficients*

[KUM 99] propose an expression for the partial coefficients. According to the authors, this expression is useful for rotating-disc contactors and for perforated-plate pulsed columns.

As regards packed columns, Laddha *et al.* [LAD 78] propose for the coefficient on the side of the dispersed phase:

$$\beta'_D = 0.02 V_{sl} Sc_D^{-0.5} \ (m.s^{-1})$$

The slip velocity V_{sl} is defined in section 1.4.8.

For the continuous phase, [SEI 88] recommend:

$$\beta'_C = 0.698\ Sc_C^{0,4}\ Re^{0,5}(1-\phi)\frac{D}{d_{32}} \quad (m.s^{-1})$$

The coefficients thus obtained are around half those that stemmed from Kumar and Hartland's relations [KUM 99], used with a power of mass agitation equal to zero. Thus, the packed column, although it is not very costly in terms of investment, is rather mediocre in comparison to the pulsed column and the rotating-disc contactor.

1.4.6. *Coalescence*

For the sake of simplicity, consider two spherical drops of similar sizes, very close to one another but without contact between them. The contiguous phase between the two drops has the form of a biconcave lens (that is, concave on both its two faces). Let us focus on the part of the lens whose diameter is, say, one quarter that of the drops. The volume of that part would be 10–50 times less than that of each drop. Let ω represent that volume.

Given the transfer of the solute, the small volume ω will very soon reach equilibrium with the drops, whilst the composition of those drops will have varied very little.

For a transfer $D \rightarrow C$, the concentration in the volume ω will very quickly reach equilibrium with the drops, and will have increased. However, the surface tension varies in the opposite direction to the concentration of the solute. It will therefore have decreased, which will cause a movement of the interface, which will move away from the axis linking the centers of the two drops, entraining a part of the volume ω. Thus, the two drops will come closer to one another and, finally, coalesce.

For a transfer $C \rightarrow D$, coalescence will be hampered, and the drops will be smaller.

1.4.7. *Marangoni effect*

The material transfer is not uniform on the surface of a drop. The compositions of the two phases at the interface, therefore, are not uniform. However, sometimes, the interfacial tension is very sensitive to the variations in composition of the two fluids. We then see movements of the interface from the zones of low interfacial tension toward the zones of high interfacial tension. These movements take place at a velocity of around one $cm.s^{-1}$ and that agitation quickly renews the concentrations, which activates the transfer which, therefore, may be up to two or three times more intense than normal (i.e. calculated with the correlations available for the chosen extractor).

It is unfortunately not possible to predict, in advance with simple resources, the Marangoni effect, which constitutes an important argument in favor of a prior study in a micropilot, i.e. with flowrates 10–50 times lower than the flowrates to be used for the industrial installation.

1.4.8. *Slip velocity*

The slip velocity, V_{sl}, is defined by:

$$V_{sl} = \frac{U_C}{1-\phi} + \frac{U_D}{\phi}$$

U_C and U_D: velocities in an empty bed, of the contiguous phase and dispersed phase respectively.

This definition is valid for the rotating-disc contactor and for the perforated-plate pulsed column.

Now let ε be the fraction of the volume of a packed column left free by the packing. We then have:

$$V_{gl} = \frac{U_C}{\varepsilon\,(1-\phi)} + \frac{U_D}{\varepsilon\phi}$$

Numerous authors have put forward expressions for the direct calculation of the slip velocity (see [KUM 99, KUM 94]). Hereinafter, though, we shall not use these expressions.

PRACTICAL NOTES.–

The time necessary for the establishment of the permanent regime is, in practice:

$$T = \frac{3\Omega}{Q_D + Q_C}$$

Ω : volume occupied by the ensemble of the two phases: m^3;

Q_C and Q_D : volume flowrates of the contiguous and dispersed phases: $m^3.s^{-1}$.

The ratio between the flowrates is, usually, such that:

$$1/3 < \frac{Q_D}{Q_C} < 3$$

1.4.9. *Flowrates on flooding (approximate method)*

We know that, by definition, the slip velocity is written:

$$V_{sl} = \frac{U_D}{\phi} + \frac{U_C}{1-\phi}$$

The velocity V_{sl} is obviously not constant when the flowrates vary. Obviously, the hypothesis was made a long time ago that we could write:

$$V_{gl} = V_0(1-\phi) \qquad [1.3]$$

Here, V_0 was considered to be constant. In addition, we accepted that, when the flowrates vary, their ratio $R = U_D/U_C$ remains constant. Therefore:

$$U_C = \frac{V_0(1-\phi)}{\frac{R}{\phi} + \frac{1}{1-\phi}}$$

When we gradually increase U_C, we reach a state where any additional flowrate does not pass through the device, and each liquid, when it is input, is entrained by the other, and thus comes out again immediately. This situation corresponds to flooding, meaning that U_C reaches a maximum. In other words:

$$\frac{dU_C}{d\phi} = 0$$

meaning that:

$$2(1-R)\phi^2 + 3R\phi - R = 0$$

The retention φ_E on engorgement is the following root of this equation:

$$\phi_E = \frac{-3R + \sqrt{R^2 + 8R}}{4(1-R)} \quad [1.4]$$

This gives us U_C and U_D.

When the correct calculation of ϕ_E is performed in iterations, it is interesting to take, as starting values, the value given by equation [1.4] above.

1.5. Dispersed phase feeder

1.5.1. *Influence of the feeder on the effectiveness of the transfer*

We use the term “feeder” for the distributor of the feed.

We define the spreading length L_h as being the horizontal distance to be covered by the dispersed phase in order to obtain a uniform distribution of this phase on the section of the extractor. Thus, for a feed by a single jet on the axis of the column, the L_h will be equal to the radius of the (circular) section of the extractor.

As long as the dispersed phase is not present on all of the section of the column, the transfer practically does not take place because the contiguous phase tends to flow where the dispersed phase is absent. Experience shows that we can express the dead height H_m as follows:

$$H_m = 15L_h$$

The usable height will be:

$$H_U = H_T - H_m$$

H_T is the total height occupied by the column's internal apparatus, i.e.:

– the packing (rings, saddles, etc.);

– the ensemble of the perforated plates in a pulsed column.

The type of feeder has no influence at all on the effectiveness of rotating-disc contactors because, from the first compartment and by the centrifugal action of the rotor's discs, the ensemble of the contiguous and dispersed phases is animated with a toroidal motion which occupies the whole section of the contactor.

1.5.2. *Proposed design for the feeder*

We can divide the section into squares (on the edges of the distributor, the "squares" are curvilinear).

In Figure 1.11, the *crosshatched surface is holed.* Each crosshatched square is fed by its own tube. Thus, there are as many tubes as there are squares, and each tube supports the square that it feeds. Using the language of botany, we say that it is an *umbel structure*. All the tubes diverge from a single vertical tube which feeds them all.

Between two crosshatched squares, there is an empty space of width 2ℓ, which is traversed by the contiguous phase, whilst the dispersed phase traverses the holes in the crosshatched surface.

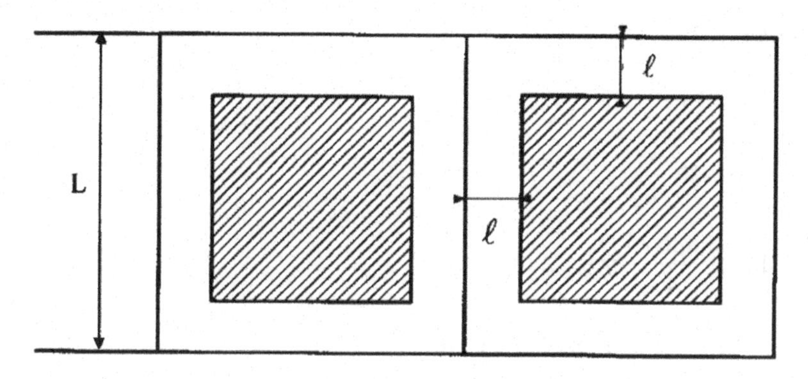

Figure 1.11. *Two contiguous squares*

The section of the column is:

$$A_{CO} = \frac{\pi D_T^2}{4} = n^2L^2 \qquad \text{where} \qquad n^2 = 4,\ 9,\ 16 \text{ or } 25$$

Thus:

$$L = \frac{D_T}{n}\sqrt{\frac{\pi}{4}} \qquad \left(\sqrt{\frac{\pi}{4}} = 0.886\right)$$

We shall choose:

$$l = \psi L = L_h$$

The section area of the extractor would be:

$$\frac{\pi D_c^2}{4} = n^2L^2 \qquad \text{whence L}$$

The total section of the crosshatched squares would have:

$$A_{ha} = n^2L^2(1-2\times\psi)^2$$

The unusable height would be:

$$H_I = 15l = 15\psi L$$

EXAMPLE 1.1.–

$$D_T = 1 \text{ m} \qquad n^2 = 16$$

$$\psi = 0.15$$

$$L = \frac{0.886 \times 1}{4} = 0{,}22 \text{ m}$$

$$L_h = 0.22 \times 0.15 = 0.033 \text{ m}$$

$$H_m = 15 \times 0.033 = 0.49 \text{ m}$$

If we had fed in the dispersed phase with a simple circle whose diameter is equal to $D_T/2$, we would have had:

$$L_h = \frac{D_T/2}{2} = 0.25 \text{ m}$$

$$H_m = 15 \times 0.25 = 3.75 \text{ m} \text{ instead of } 0.49 \text{ m}.$$

NOTE.–

Often, the diameter of packed columns or perforated-plate pulsed columns is no more than 0.3 m, which dispenses with the installation of sophisticated feeders such as that described above.

1.5.3. *Velocity through the holes in the distributor*

For a low flowrate, only a limited number of holes operate and are therefore active. They work drop by drop. When the flowrate is increased, the number of active holes increases and, when they are all working, a slight increase in flowrate results in the appearance of jets at the outlet from the holes. The velocity through the holes is therefore equal to V_j.

The critical velocity V_{cr} through the holes is reached when the length of the jets is maximal. At this velocity, the jets resolve into drops, all of which have the same diameter d_{32}, and that diameter is minimum because if we increase the velocity further, the particle-size distribution of the drops spreads, i.e. there is the simultaneous appearance of numerous fine drops of diameter d_m and large drops of diameter d_M such that:

$$d_m < d_{32} < d_M$$

In practical terms, to calculate the critical velocity, we use the correlation put forward by Skelland and Huang [SKE 79]. We take a hole diameter d_o of between 1 and 3 mm, and calculate the Eötvös number:

$$Eö = \frac{d_o^2 \Delta\rho g}{\sigma}$$

$$\text{If } Eö < 0.615 \quad d_{cr} = \frac{2.07\ d_o}{1 + 0.485\ Eö}$$

$$\text{If } Eö > 0.615 \quad d_{cr} = \frac{2.07\ d_o}{0.12 + 1.51\ Eö^{0.5}}$$

Hence:

$$V_{cr} = 0.9\left(\frac{d_{cr}}{d_o}\right)^2 \left[\frac{\sigma}{d_{cr}(0.5137\rho_D + 0.4719\rho_C)}\right]^{0.5}$$

EXAMPLE 1.2.–

$$d_o = 0.002 \text{ m}$$

$$Eö = \frac{0.002^2 \times 200 \times 9.81}{0.015} = 0.5232 < 0.615$$

$$d_{cr} = \frac{2.07 \times 0.002}{1 + 0.485 \times 0.5232} = 0.00330 \text{ m}$$

$$V_{cr} = 0.9\left(\frac{3.3}{2}\right)^2 \left[\frac{0.015}{0.0033(0.5137 \times 900 + 0.4719 \times 1100)}\right]^{0.5}$$

$$V_{cr} = 0.167 \text{ m.s}^{-1}$$

1.5.4. *Aperture of the distributor*

Let ϕ be the ratio of the surface of the holes in a crosshatched square to the total surface of that square. The parameter ϕ is the "aperture" of the crosshatched surface.

The velocity of the dispersed phase through the holes will be taken as equal to the critical velocity:

$$V_{cr} = \frac{Q_D}{\phi n^2 L^2 (1-2\psi)^2} = \frac{U_D}{\phi(1-2\psi)^2}$$

Thus:

$$\phi = \frac{U_D}{V_{cr}(1-2\psi)^2}$$

EXAMPLE 1.3.–

$$U_D = 0.0025 \text{ m.s}^{-1} \qquad \psi = 0.2 \qquad V_{cr} = 0.167 \text{ m.s}^{-1}$$

$$\phi = \frac{0.0025}{0.167 \times (1 - 2 \times 0.2)^2} = 0.0416$$

NOTE.–

It is costly to create such a distributor, but it is also possible to use a tube in the form of a horizontal spiral and pierced with holes in the part facing into the column.

An intermediary solution would contain several concentric horizontal circles, fed separately.

Let p be the distance between the axes of two consecutive spires of the spiral or between the (circular) axes of two contiguous circles, and d_T the external diameter of the tube used to make the spiral or circles.

We can write:

$$L_h = \frac{1}{2}(p - d_T) \quad \text{and} \quad H_m = 7.5(p - d_T)$$

1.5.5. *Step between the holes*

Generally, the holes are arranged in an equilateral triangular pattern.

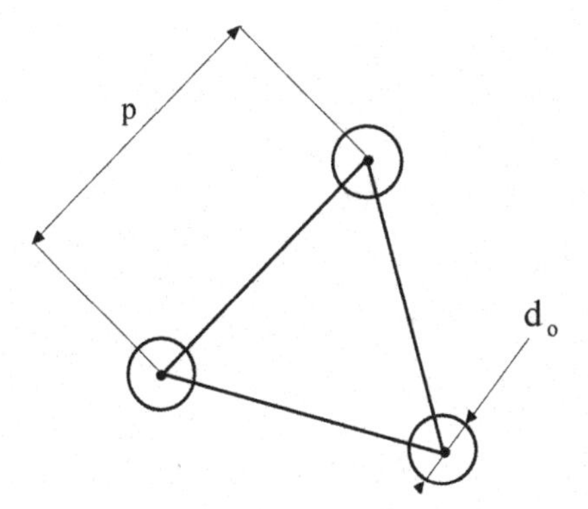

Figure 1.12. *Triangular pattern*

The area of a triangle is:

$$S_o = p^2 \times \frac{\sqrt{3}}{2} \times \frac{1}{2} = \frac{\sqrt{3}}{4} p^2$$

In the triangle, the area occupied by the holes is:

$$S_1 = \frac{1}{2}\left(\frac{\pi d_o^2}{4}\right) = \frac{\pi}{8} d_o^2$$

By definition, the fraction of surface available to the liquid through the holes is written:

$$\phi = S_1 / S_o = \frac{4\,\pi}{8\,\sqrt{3}} \frac{d_o^2}{p^2}$$

Therefore:

$$p = d_o \left(\frac{\pi}{2\sqrt{3}\,\phi} \right)^{1/2}$$

EXAMPLE 1.4.–

$$\phi = 0.0416$$

$$p = d_o \left(\frac{\pi}{2\sqrt{3} \times 0{,}0416} \right)^{1/2}$$

$$p = 4.67\, d_o$$

With spacing of $p - d_o = 3.67\, d_o$, there is a low risk of parasitic interactions between the holes.

1.6. Conclusion – pilot studies

1.6.1. *Advantages and disadvantages of different types of extractors*

The simplest extractor is obviously the "empty" column with no internal elements. Unfortunately, the equivalence of such a column is generally situated between 1 and 1.5 theoretical stages. Indeed, the drops of dispersed phase generally flow in the vicinity of the axis and the continuous phase channeled in the vicinity of the wall. All of this greatly decreases the effective surface area for transfer between the two phases.

To avoid these drawbacks, one envisaged solution is to place packing inside the column. Indeed, the equivalence of such a column can reach 3 or 4 theoretical stages with conventional packing (rings or saddles). However, there are still preferential passages (channeling) for each of the two liquids. In addition, the size of drops may reach up to 3 or 4 mm.

The pulsed column with perforated plates does not create channeling. Furthermore, the alternating motion between the holes maintains a small diameter for the drops and activates material transfer between the two liquids. The result of this is that the equivalence of these columns is up to two times greater than that of conventional packed columns (obviously for the same height of the column). However, we must beware of the clogging of certain holes if there is interfacial fouling.

Packed columns and pulsed columns share the drawback of low radial dispersion, which means that if, in a given section, the liquids entering into the column are poorly distributed, they will remain so over a not-insignificant height. Thus, the distributor needs to be carefully designed.

The rotating-disc extractor does not exhibit this latter disadvantage. In addition, the influence of the discs maintains a low value of the size of the drops. Finally, this device is not vulnerable to interfacial fouling.

Serial mixers-decanters may be of interest, in spite of their horizontal bulk. An array of mixers may be equivalent to 10 theoretical stages.

When the difference between the densities of the two phases is less than 50 kg.m^{-3}, the centrifugal extractor may be a good solution.

1.6.2. *Equivalence of devices*

Type of device	Equivalence (in theoretical stages)
Spray column (without internal equipment)	1–1.5
Packed column	3 or 4
Pulsed column with perforated plates	5–7
Rotating-disc contactor	4–6
Centrifugal extractor	1–2
Mixers/decanters (a single device)	0.8–1

Table 1.1. *Equivalence of extractors*

Naturally, these equivalences are merely approximate.

1.6.3. *Choice of the dispersed phase*

The exchange surface is larger if we disperse the liquid which has the greater flowrate. However, this criterion is limited if a phase inversion occurs.

The intensity of the transfer may depend on its direction. In the view of some, it would be preferable for the solute to pass from the continuous phase into the dispersed phase for reasons of non-coalescence (see section 1.4.6).

1.6.4. *Use of pilots in liquid–liquid extraction*

The use of packed columns is becoming rare. Indeed, the behavior of two liquids is *unpredictable* if the normal size of the packing is less than 1 cm. However, at that size, the diameter of the column must be at least 10 cm, which leads to a not-insignificant flowrate. In addition, if the liquids are valuable chemical species, the micropilot would be almost as costly as the industrial column, and also carry with it the hazards pertaining to the distribution system for the discontinuous phase.

It is easier to choose a pulsed column with perforated plates. The extrapolation will be correct if, in the pilot and the industrial installation, we preserve:

– the mass agitation power Ψ;

– the diameter of the holes;

– the aperture of the plates (holed fraction of surface);

and finally, if the horizontal length L_h characterizing the distributor is known precisely in the pilot and in the industrial arena.

However, it proves to be even easier to choose a rotating-disc contactor, using the dimensions specified by Kumar *et al.* [KUM 99].

We then simply need to preserve the mass agitation power Ψ to conserve the value of the transfer coefficient. The design of the distributor is unimportant.

Thus, we can see that the rotating disc device needs to be used in most cases – especially if there is a chemical reaction at the interface.

However, in the nuclear industry, it is preferable to avoid the presence of moving mechanical parts in direct contact with radioactive liquids, and therefore preference is given to the pulsed column with perforated plates, because it is then possible to install the pulsation system (e.g. a volumetric diaphragm pump without valves) at a certain distance from the extractor. Arrays of mixers/decanters are also commonly used in this industry.

1.6.5. *Additional reasons to use a pilot*

There are two types of such reasons, pertaining to the presence of:

– the Marangoni effect (unpredictable), which intensifies transfer by action on the overall material transfer coefficient,

– coalescence between drops, which decreases the interfacial area,

– a reaction at the interface.

2

Three Examples of Calculation for a Differential Liquid–Liquid Extractor

2.1. General

2.1.1. *Interfacial tension*

It is easy to measure the interfacial tension using a ring or blade tensiometer.

It can also be estimated by using Treybal's [TRE 63] correlation. We calculate:

$$Y = X_{EO} + X_{OE} + \frac{1}{2}(X_{TE} + X_{TO})$$

The X are the molar fractions:

X_{EO} : water in the organic phase;

X_{OE} : organic solvent in the aqueous phase;

X_{TE} and X_{TO} : solute to be transferred in the aqueous phase and in the organic phase.

If $0.0004 < Y < 0.4$ $\qquad \sigma = 17 \log_{10}(1/Y) - 5$

If $0.4 < Y < 1$ $\qquad \sigma = 4.43 \log_{10}(1/Y)$

The interfacial tension σ is expressed in dyne.cm^{-1}, where:

$$1\ \text{dyne.cm}^{-1} = 0.001\ \text{N.m}^{-1}$$

2.1.2. *Physical properties and notations*

Here, let us recap the conventional notations and SI units which we shall employ:

σ : interfacial tension: N.m^{-1};

g : acceleration due to gravity: 9.81 m.s^{-2};

ρ_C, ρ_D : densities of the continuous- and dispersed phases: kg.m^{-3};

$\Delta\rho$: $|\rho_C - \rho_D|$: kg.m^{-3};

V_C, V_D: velocities in an empty bed of the continuous- and dispersed phases: m.s^{-1};

V_{CE}, V_{DE} : velocities of the same phases on flooding: m.s^{-1};

μ_C, μ_D: dynamic viscosities of the continuous- and dispersed phases: Pa.s;

D_C, D_D: diffusivities of the solute in the continuous- and dispersed phases: m^2.s^{-1};

Sc_C, Sc_D: Schmidt numbers of the continuous- and dispersed phases: $Sc = \dfrac{\mu}{\rho D}$;

m : slope of the distribution curve: dx_D^* / dx_C;

K_{OC}: overall transfer coefficient expressed in relation to the continuous phase: s^{-1};

D_{AC}, D_{AD}: axial dispersion coefficients (dispersivity) in the continuous- and dispersed phases: m^2.s^{-1};

a: interfacial area of transfer (m^2 per m^3 of column): m^{-1}.

IMPORTANT NOTE.–

For each differential extractor, we shall give a numerical example. In these calculations, we shall systematically use the following values:

$$\sigma = 0.015\ \mathrm{N.m^{-1}} \qquad m = 0.7$$

$$\mu_C = 1.5.10^{-3}\ \mathrm{Pa.s} \qquad \mu_D = 2.10^{-3}\ \mathrm{Pa.s}$$

$$\rho_C = 1100\ \mathrm{kg.m^{-3}} \qquad \rho_D = 900\ \mathrm{kg.m^{-3}}$$

$$D_C = 1.2.10^{-9}\ \mathrm{m^2.s^{-1}} \quad D_D = 2.5.10^{-9}\ \mathrm{m^2.s^{-1}}$$

$$R = V_D / V_C = 1{,}3$$

The transfer takes place from the continuous phase toward the dispersed phase.

2.2. Packed columns

2.2.1. *Constraints on the nominal size of the packing*

By considerations which we shall not detail here, Gayler *et al.* [GAY 53] showed that the distance traveled by a drop between two collisions with the packing is:

$$\overline{s} = 0.38 d_N - 0.92 \left(\frac{\sigma}{g \Delta \rho} \right)^{1/2} \qquad \text{(S.I.)}$$

d_N : nominal size of the packing: m

The types of packing considered are rings (Raschig or Pall) and saddles (Berl or Intalox). The appendix to this book gives the characteristics of these packings.

The distance $\overline{s}$ becomes zero for:

$$d_N^* = 2{,}42 \left(\frac{\sigma}{g \Delta \rho} \right)^{1/2} \qquad \text{(S.I.)}$$

If the packing is such that $d_N > d_N^*$, the drops continue to move, but at the cost of a deformation, so the laws established for sizes $d_N < d_N^*$ are no longer valid.

In addition, in order for the column to work without instability, and for the porosity (empty fraction) of the packing to be sufficiently uniform across the section of the device, we must have:

$$d_N < \frac{D_T}{10}$$

D_T : diameter of the column: m.

Indeed, near to the wall, the porosity of the packing is always greater than it is at the heart of the column.

EXAMPLE 2.1.–

$$\sigma = 0.015 \text{ kg.s}^{-2} \qquad \Delta\rho = 200 \text{ kg.m}^{-3}$$

$$D_T = 0.10 \text{ m}$$

$$d_N^* = 2.42\left(\frac{0.015}{9.81 \times 200}\right)^{1/2} = 0.0067\text{m}$$

Therefore, we must have:

$$0.010 \text{ m} \geq D_N \geq 0.0067$$

1 cm packing should be acceptable.

2.2.2. *Flooding [KUM 94]*

On flooding, the velocity in an empty bed of the contiguous phase is:

$$V_{CE} = \frac{\sqrt{g/a_T}}{\left(1 + R^{1/2}\right)^2} C_I \varepsilon^{1.54} \left(\frac{\Delta\rho}{\rho_D}\right)^{0.41} \times \left[\frac{1}{a_T}\left(\frac{\Delta\rho^2 g}{\mu_C^2}\right)^{1/3}\right]^{0.3} \times \left[\mu_C \sqrt{\frac{a_T}{\Delta\rho\sigma}}\right]^{0.15}$$

Nature of the packing	Value of C_1
Raschig rings and Intalox saddles	0.28
Berl saddles	0.37
Pall rings	0.20

a_T : surface of the packing expressed in relation to the volume of the column: m^{-1};

R : ratio of the volume flowrates of the dispersed phase to the continuous phase:

$$R = V_D / V_C$$

EXAMPLE 2.2.–

25 mm Raschig rings:

$$a_T = 200\ m^{-1} \qquad R = 1.3 \qquad \varepsilon = 0.73$$

$$V_{CE} = \frac{\sqrt{9.81/200}}{\left(1+1.3^{1/2}\right)^2} 0.28 \times 0.73^{1.54} \left(\frac{200}{900}\right)^{0.41} \times \left[\frac{1}{200}\left(\frac{200^2 \times 9.81}{\left(1.5.10^{-3}\right)^2}\right)^{1/3}\right]^{0.3}$$

$$\times \left[1.5.10^{-3}\sqrt{\frac{200}{200 \times 0.015}}\right]^{0.15}$$

$$V_{CE} = 0.006269\ m.s^{-1}$$

Let us adopt a rate of flooding of 50%:

$$V_C = 0.5 \times 0.006269 = 0.003134\ m.s^{-1}$$

$$V_D = 1.3 \times 0.003134 = 0.004075\ m.s^{-1}$$

2.2.3. *Retention of dispersed phase [KUM 94]*

The retention is expressed by:

$$\phi = C_1 \varepsilon^{-1.11} \left(\frac{\Delta\rho}{\rho_C}\right)^{-0.5} \left[\frac{1}{a_T}\left(\frac{\rho_C g}{\mu_C^2}\right)^{1/3}\right]^{-0.72} \times \left(\frac{\mu_D}{\mu_C}\right)^{0.1} \left[V_D \left(\frac{\rho_C}{g\mu_C}\right)^{1/3}\right]^{1.03} \times$$

$$\exp\left[0.95\,V_C\left(\frac{\rho_C}{g\mu_C}\right)^{1/3}\right]$$

The coefficient C_1 depends on the direction of the transfer:

$$C \to D : C_1 = 6.16$$

$$D \to C : C_1 = 3.76$$

EXAMPLE 2.3 (cont.).–

$$\phi = 6.16 \times 0.73^{-1.11} \times \left(\frac{200}{1100}\right)^{-0.5} \times \left[\frac{1}{200}\left(\frac{1100^2 \times 9.81}{(1.5.10^{-3})^2}\right)^{1/3}\right]^{-0.72} \times \left(\frac{2}{1.5}\right)^{0.1} \times$$

$$\left[0.004075\left(\frac{1100}{9.81 \times 0.0015}\right)^{1/3}\right]^{1.03} \times \exp\left[0.95 \times 0.003134\left(\frac{1100}{9.81 \times 0.0015}\right)^{1/3}\right]$$

$$\phi = 0.156$$

2.2.4. *Mean drop diameter [KUM 94]*

The mean diameter d_{32} is obtained by:

$$d_{32} = \left(\frac{\sigma}{\Delta\rho g}\right)^{1/2} C_1 \left[\frac{\mu_e g^{1/4}}{\Delta\rho^{1/4}\sigma^{3/4}} \frac{\rho_C}{\rho_D}\right]^{0.19}$$

Depending on the direction of the transfer:

$$C \to D : C_1 = 2.24$$

$$D \to C : C_1 = 3.13$$

μ_C and ρ_C: viscosity and density of water: Pa.s and kg.m^{-3}.

EXAMPLE 2.4.–

$$C \to D : d_{32} = \left(\frac{0.015}{9.81 \times 200}\right)^{1/2} \times 2.24\left(\frac{10^{-3} \times 9.81^{0.25}}{200^{0.25} \times 0.015^{0.75}} \times \frac{1000}{900}\right)^{0.19}$$

$$d_{32} = 0.0026\text{ m}$$

2.2.5. *Slip velocity [KUM 94]*

We calculate:

$$V_{gl} = \left(\frac{g\mu_C}{\rho_C}\right)^{1/3} C_1 \varepsilon^{-0.17} \times \left(\frac{\Delta\rho}{\rho_C}\right)^{-0.41} \left[\frac{1}{a_T}\left(\frac{\rho_C^2 g}{\mu_C^2}\right)^{1/3}\right]^{0.59} \times \left(\frac{\mu_D}{\mu_C}\right)^{-0.1} (1-\phi)$$

Depending on the direction of the transfer:

$$C \rightarrow D : C_1 = 0.27$$

$$D \rightarrow C : C_1 = 0.38$$

EXAMPLE 2.5.–

$$V_{sl} = \left(\frac{9.81 \times 0.0015}{1100}\right)^{1/3} \times 0.27 \times 0.73^{-0.17} \times \left(\frac{\Delta\rho}{\rho_C}\right)^{0.41} \left[\frac{1}{200}\left(\frac{1100^2 \times 9.81}{0.0015^2}\right)^{1/3}\right]^{0.59}$$

$$\times \left(\frac{2}{1.5}\right)^{-0.1} \times 0.844$$

$$V_{gl} = 0.0384 \text{ m.s}^{-1}$$

NOTE.–

We must verify that:

$$V_{sl} = \frac{V_C}{\varepsilon(1-\phi)} + \frac{V_D}{\varepsilon\phi}$$

Thus:

$$0.0384 = \frac{0.004075}{0.73 \times 0.156} + \frac{0.003134}{0.73(1 - 0.156)} = 0.04086$$

We can accept the arithmetic mean:

$$V_{sl} = \frac{1}{2}(0.0384 + 0.04086) = 0.03973 \text{ m.s}^{-1}$$

2.2.6. *Transfer coefficients*

According to Laddha and Degaleesan [LAD 78]:

$$\beta_D = 0.023 V_{gl} \, Sc_D^{-0.5}$$

V_{sl} : slip velocity: m.s^{-1};

Sc_D : Schmidt number for the dispersed phase;

D_D : diffusivity of the solute transferred in the dispersed phase: $\text{m}^2.\text{s}^{-1}$ according to Seibert and Fair [SEI 88]

$$Sh_C = 0.698 Sc_C^{0.4} Re^{0.5}(1 - \phi)$$

Sh_D : Sherwood number

$$Sh_C = \frac{\beta_C d_{32}}{D_C}$$

Re : Reynolds number

$$Re = \frac{\rho_C d_{32} V_{gl}}{\mu_C}$$

ϕ : retention of the dispersed phase.

Hence:

$$\beta_C = \frac{Sh_C D_C}{d_{32}}$$

EXAMPLE 2.6.–

$$V_{gl} = 0.03973 \text{ m.s}^{-1} \qquad d_{32} = 0.0026 \text{ m}$$

$$Sc_D = \frac{2.10^{-3}}{900 \times 2.5.10^{-9}} = 889 \quad \phi = 0.156$$

$$\beta_D = 0.023 \times 0.03973 \times 889^{-0.5}$$

$$\beta_D = 3.06.10^{-5} \text{ m.s}^{-1}$$

$$Re = \frac{1100 \times 0.0026 \times 0.03973}{0.0015} = 75.75$$

$$Sc_C = \frac{1.5.10^{-3}}{1100 \times 1.2.10^{-9}} = 1136$$

$$Sh_C = 0.698 \times 1.136^{0.4} \times 75.75^{0.5} (1 - 0.156) = 85.49$$

$$\beta_C = \frac{85.49 \times 1.2.10^{-9}}{0.0026} = 3.95.10^{-5} \text{ m.s}^{-1}$$

2.2.7. *Axial dispersion coefficients [VER 66]*

Here we shall put forward analytical expressions to show the figures found by the various authors:

1) Dispersed phase:

$$Pe_D = 0.45 \exp\left[-\frac{0.405 \, V_C}{V_D} \right]$$

$$D_{AD} = \frac{V_D d_N}{Pe_D}$$

d_N: nominal size of the packing: m;

V_D and V_C: velocities in an empty bed of the dispersed- and contiguous phases: $m.s^{-1}$;

D_{AD} : axial dispersion coefficient for the dispersed phase: $m^2.s^{-1}$.

2) Continuous phase:

We calculate:

$$X = \frac{V_D}{V_C}\left[\frac{0.6\mu_C}{\rho_C d_N V_C}\right]^{0.5}$$

$$Pe_C = 0.01268X^{-0.9043}$$

We must verify that:

$$Pe_C \leq 0.1443 \text{Ln Re} - 0.232 \qquad [2.1]$$

where:

$$Re = \frac{V_C \rho_C d_N}{\mu_C \ (1-\varepsilon)}$$

If this inequality is not satisfied, then for Pe_C, we take the value of the right-hand side of inequality [2.1].

ε : porosity of the packing: empty relative to the volume of the column

and, finally:

$$D_{AC} = \frac{V_C d_N \varepsilon}{Pe_C}$$

These calculations are valid for rings (Raschig or Pall) and saddles (Berl or Intalox).

EXAMPLE 2.7.–

$$V_D = 0.004075\ m.s^{-1} \qquad d_N = 0.025\ m$$

$$V_C = 0.003134\ m.s^{-1} \qquad \varepsilon = 0.73\ (\text{Raschig rings})$$

$$Pe_D = 0.45 \exp\left[-\frac{0.405 \times 0.003134}{0.004075}\right] = 0.3114$$

$$D_{AD} = \frac{0.004075 \times 0.025}{0.3114}$$

$$D_{AD} = 3.27.10^{-4}\ m^2.s^{-1}$$

$$X = \frac{0.004075}{0.003134}\left(\frac{0.6 \times 0.0015}{0.025 \times 1100 \times 0.003134}\right)^{0.5} = 0.133$$

$$Pe_C = 0.01268 \times 0.133^{-0.9043} = 0.0786$$

$$Re = \frac{0.003134 \times 1100 \times 0.025}{0.0015 \times (1 - 0.73)} = 212.80$$

We can verify that:

$$0.0786 < 0.1443 \mathrm{Ln} 212.80 - 0.232 = 0.5415$$

Thus:

$$D_{AC} = \frac{0.003134 \times 0.73 \times 0.025}{0.0786}$$

$$D_{AC} = 7.28.10^{-4}\ m^2.s^{-1}$$

2.3. Rotating-disc contactor

2.3.1 *Presentation and advantage of the device*

The liquid medium is animated with a toroidal motion, and rolls up in a helix around each tore. By that circulation, the medium becomes homogeneous over a given section; it is therefore pointless to use sophisticated distributors and the feed takes place advantageously by a circular tube pierced through with holes in the part facing the interior of the device. The diameter of the axial circle of the tube will be:

$$D_A = \frac{1}{2}(D_R + D_S)$$

The rotating-disc contactor is not sensitive to the presence of fouling at the interface, or to the presence of solid particles in the continuous phase.

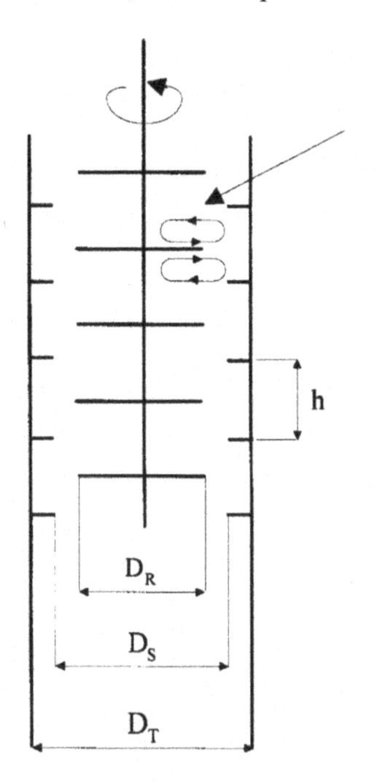

Toric circulation (2 tores in each compartment)

h: height of a compartment

D_R: diameter of the discs of the rotor

D_S: internal diameter of the coronas of the stator

D_T: diameter of the “tower” – i.e. the contactor

Figure 2.1. *Principle behind the rotating-disc contactor*

2.3.2. *Definition of the rotating-disc contactor*

The successive steps, we believe, are as follows:

1) choice of the diameter for a pilot;

2) in view of the designing standards, definition of the geometry of the pilot;

3) on the basis of a peripheral velocity of the rotor between 1 and 2 $m.s^{-1}$, definition of the rotation speed of the rotor of the pilot;

4) calculation of the gravimetric agitation power of the pilot;

5) based on a velocity in an empty bed of the contiguous phase equal to 0.0015 $m.s^{-1}$, estimation of the diameter of the contactor and the height h of the compartments;

6) mean drop diameter, retention on engorgement;

7) velocities in an empty bed on engorgement, diameter and, in accordance with the design standards, geometry of the device;

8) retention of the dispersed phase by the extractor and rotation speed of its rotor for a gravimetric agitation power equal to that of the pilot.

2.3.3. *Design standards [KUM 99a]*

If D_T is the diameter of the device, we have:

$$D_R/D_T = 0.6 \qquad D_S/D_T = 0.7$$

$$h = 0.13\, D_T^{0.67} \qquad e = 0.49$$

D_R : diameter of the rotor (diameter of a disc): m;

D_S : internal diameter of the coronas: m;

h : height of a compartment: m.

In order to obtain an initial estimation of the geometry of the contactor, we shall write:

$$V_C = 0.0015\,m.s^{-1}$$

$$D_T = \sqrt{\frac{4Q_C}{\pi V_C}}$$

V_C : velocity in an empty bed of the contiguous phase: $m.s^{-1}$;

Q_C : flowrate of contiguous phase: $m^3.s^{-1}$.

EXAMPLE 2.8 (Geometry of the pilot).–

We have chosen:

$$D_T = 0.07\ m$$

Hence:

$$D_S = 0.07 \times 0.7 = 0.049\,m$$

$$D_R = 0.07 \times 0.6 = 0.042\ m \qquad h = 0.13 \times 0.07^{0.67} = 0.0219\,m$$

EXAMPLE 2.9 (Industrial unit).–

$$Q_C = 0.001178\ m^3.s^{-1} \qquad V_C = 0.0015\ m.s^{-1}.$$

Thus:

$$D_T = \sqrt{\frac{4 \times 1.178 \times 10^{-3}}{\pi \times 0.0015}} = 1\ m$$

$$D_R = 0.6\ m \qquad D_S = 0.7\ m$$

$$h = 0.13 \times 1^{0.67}$$

$$= 0.13\ m$$

2.3.4. *Rotation speed of the pilot*

Rather than using Kalaichelvi's [KAL 97] method, which gives a critical velocity which we then increase by 30%, we shall employ the concept of the peripheral velocity of the rotor (of the discs) and we accept a value of $1-2\ m.s^{-1}$ for that velocity.

EXAMPLE 2.10.–

$D_R = 0.042\ m \qquad V_{peri\ rot} = 2\ m.s^{-1}$

$$N = \frac{2}{\pi \times 0.042} = 15.16\ rev.s^{-1}$$

2.3.5. *Gravimetric agitation power (pilot) [KUM 96]*

We successively calculate:

$$Re = \frac{N\ D_R^2\ \rho_C}{\mu_C}$$

$$N_P = \frac{109.36}{Re} + 0.74 \left[\frac{1000 + 1.2\ Re^{0.72}}{1000 + 3.2\ Re^{0.78}}\right]^{3.30}$$

$$P = N_P\ N^3\ D_R^5\ \rho_C$$

The gravimetric agitation power is:

$$\psi = \frac{4P}{\pi\ D_T^2\ h \rho_C}$$

N_P : power number;

N: rotation speed: $rev.s^{-1}$;

h : height of a compartment: m;

P : power developed by a disc: W;

ψ : mass agitation power: $W.kg^{-1}$;

D_R : diameter of the rotor (diameter of a disc): m;

Re : Reynolds number.

EXAMPLE 2.11 (Pilot).–

$$D_T = 0.07 \text{ m} \quad D_R = 0.042 \text{ m} \qquad h = 0.13 \times 0.07^{0.67} = 0.0219 \text{ m}$$

$$Re = \frac{15.16 \times 0.042^2 \times 1100}{1.5.10^{-3}} = 19611$$

$$N_P = \frac{109.36}{19611} + 0.74\left[\frac{10^3 + 1.2 \times 19611^{0.72}}{10^3 + 3.2 \times 19611^{0.72}}\right]^{3.3} = 0.1654$$

$$P = 0.1654 \times 15.16^3 \times 0.042^5 \times 1100 = 0.083$$

$$\psi = \frac{4 \times 0.083}{\pi \times 0.07^2 \times 0.0219 \times 1100} = 0.895 \text{ W.kg}^{-1}$$

2.3.6. *Mean drop diameter*

According to Kumar and Hartland [KUM 99]:

$$d_{32} = \frac{0{,}611Ch}{A + B}$$

C is equal to: 1 for the transfer $C \to D$

1.29 for the transfer $D \to C$

$$A = 2.54\left(\frac{\sigma}{\Delta\rho g h^2}\right)^{1/2}$$

$$B = \frac{1}{0.97}\left[\frac{g}{\psi}\left(\frac{g\sigma}{\rho}\right)^{1/4}\right]^{-0.45} \times \left[\frac{1}{h}\left(\frac{\sigma}{\rho_C g}\right)^{1/2}\right]^{-1.12}$$

EXAMPLE 2.12.–

$$h = 0.13\text{ m} \qquad C = 1$$

$$A = 2.54\left(\frac{0,015}{200\times 9.81\times 0.13^2}\right)^{1/2} = 0.05402$$

$$B = \frac{1}{0.97}\left[\frac{9.81}{0.895}\left(\frac{9.81\times 0.015}{1100}\right)^{1/4}\right]^{-0.45} \times \left[\frac{1}{0.13}\left(\frac{0.015}{9.81\times 1100}\right)^{1/2}\right]^{-1.12} = 185.59$$

$$d_{32} = \frac{0.611\times 0.13}{0.05402 + 185.59} = 0.42.10^{-3}\text{m}$$

NOTE.–

As term A is much smaller than B, we can safely neglect it, and we obtain:

$$d_{32} = 0.611\times C\times 0.97\left[\frac{g}{\psi}\left(\frac{g\sigma}{\rho_C}\right)^{1/4}\right]^{0.45} \times \left[\frac{1}{h}\left(\frac{\sigma}{\rho_C g}\right)^{1/2}\right]^{1.12} \times h$$

Note that d_{32} varies with $h^{-0.12}$.

2.3.7. *Hold-up on flooding*

According to Kumar and Hartland [KUM 88], we calculate:

$$\beta = \frac{24\mu_C}{0.53 d_{32}\,\rho_C} \qquad \text{and} \qquad \gamma = \frac{4 d_{32} g \Delta\rho}{1.59\rho_C}$$

$$B = (R-1)\phi_E^2 - 2R\phi_E + R$$

$$\Gamma = 1 + 4.56\ \phi_E^{0.73}$$

$$\delta = \left[\beta^2 + \frac{4\gamma(1-\phi_E)}{\Gamma}\right]^{0.5}$$

R: ratio V_D/V_C of the flowrates of the dispersed and continuous phases

We need to find the value of ϕ_E which yields a zero value for the function:

$$F(\phi_E) = (\delta - \beta)B - \frac{2\gamma\phi_E}{\Gamma}(B + \phi_E)\left[1 + \frac{3.33\phi_E^{-0.27}(1-\phi_E)}{\Gamma}\right]$$

ϕ_E is the retention on engorgement.

The calculation takes place in iterations, giving ϕ_E the initial value:

$$\phi_E^{(0)} = \frac{-3R + \sqrt{R^2 + 8R}}{4(1-R)}$$

EXAMPLE 2.13.–

$$R = 1.3$$

$$\beta = \frac{24 \times 1.5.10^{-3}}{0.53 \times 0.42.10^{-3} \times 1100} = 0.1470$$

$$\gamma = \frac{4 \times 0.42.10^{-3} \times 9.81 \times 200}{1.59 \times 1100} = 0.001884$$

$$\phi_E^{(0)} = \frac{-3 \times 1.3 + \sqrt{1.3^2 + 8 \times 1.3}}{4(1-1.3)} = 0.35$$

The calculations give:

$$F(0.35) = 0.001609$$

$$F(0.38) = 0.001160$$

$$F(0.42) = 0.00058$$

We shall adopt:

$$\phi_E = 0.42$$

2.3.8. *Diameter of the extractor [KUM 99a]*

According to Kumar and Hartland [KUM 88], the velocity in an empty bed of the continuous phase on flooding is written:

$$V_{CE} = \frac{(\delta - \beta)\,\phi_E\,(1 - \phi_E)}{2\,(\phi_E + R - R\phi_E)}$$

We place ourselves at 50% of the flooding:

$$V_C = 0.5 V_{CE} \qquad \text{and} \qquad V_D = R V_C$$

EXAMPLE 2.14.–

$$V_{CE} = \frac{(0.1513 - 0.1470) \times 0.42 \times 0.58}{2\,(0.42 + 1.3 - 0.546)}$$

$$V_{CE} = 0.4461.10^{-3} \text{m.s}^{-1}$$

$$V_C = 0.5 \times 0.4461.10^{-3} = 0.223.10^{-3} \text{m.s}^{-1}$$

$$V_D = 1.3 \times 0.223.10^{-3} = 0.29.10^{-3} \text{m.s}^{-1}$$

$$D_T = \sqrt{\frac{4 \times 0.001178}{\pi \times 0.223.10^{-3}}}$$

$$D_T = 2.59 \text{ m}$$

Thus:

$$h = 0.13 \times 2.59^{0.67} = 0.2462 \text{ m}$$

We had supposed that in the industrial unit:

$$h = 0.13 \text{ m}$$

We therefore need to correct the diameter of the drops:

$$d_{32} = 0.42.10^{-3} \times \left(\frac{0.246}{0.13} \right)^{-0.12} = 0.39.10^{-3} \text{ m}$$

This drop size is a little too small. The most common diameter is 10^{-3} to $1.5.10^{-3}$. Therefore, we need to reduce the speed of rotation of the discs of the pilot, which accounts for the advantage of a velocity adjuster on the motor of the rotor. Furthermore, the diameter of 2.59 m may be reduced a little further by increasing the size of the drops, i.e. by *decreasing the rotation speed of the pilot* and, consequently, the mass power accepted throughout this study.

2.3.9. *Hold-up of dispersed phase*

According to Kumar and Hartland [KUM 99], the hold-up is expressed by:

$$\phi = \left\{ 0.32 + \left[\frac{\psi}{g} \left(\frac{\rho_C}{g\sigma} \right)^{1/4} \right]^{0.71} \right\} \left[V_D \left(\frac{\rho_C}{g\sigma} \right)^{1/4} \right]^{0.75} \times \exp\left[7.3 V_C \left(\frac{\rho_C}{g\sigma} \right)^{1/4} \right] \times$$

$$\left[d_{32} \left(\frac{\rho_C g}{\sigma} \right)^{1/2} \right]^{-0.26} \left(\frac{\Delta\rho}{\rho_C} \right)^{-0.67} \left(\frac{\mu_D}{\mu_C} \right)^{0.14} \times \left(\frac{D_R}{h} \right)^{0.62} \times 0.49^{-0.21} \times \left[h \left(\frac{\rho_C g}{\sigma} \right)^{1/2} \right]^{-0.1}$$

EXAMPLE 2.15.–

$$V_C = 0.223.10^{-3} \text{m.s}^{-1} \quad \psi = 0.895 \text{ W.kg}^{-1} \quad D_R = 1.554 \text{ m}$$

$$V_D = 0.29.10^{-3} \text{m.s}^{-1} \quad d_{32} = 0.39.10^{-3} \text{m} \quad h = 0.2462$$

$$\phi = 1.2097 \times 0.011833 \times 1.01525 \times 1.3332 \times 3.1336$$
$$\times 1.1019 \times 1.1616 \times 3.134 \times 0.5877$$

$$\phi = 0.143$$

2.3.10. *Speed of rotation for the industrial device*

Suppose that we know the dimensions of the industrial device and its gravimetric agitation power (which is equal to that of the pilot). We have:

$$P = \frac{\pi}{4}\,\psi\, D_T^2\, h\rho_C \qquad \text{and} \qquad N_P N^3 = \frac{P}{D_R^5 \rho_C}$$

$$\mathrm{Re} = \left(\frac{D_R^2\, \rho_C}{\mu_C}\right) N$$

Very generally, the velocity N is no greater than 1 rev.s^{-1}, which means that:

– the Reynolds number is greater than 10^5;

– the expression of N_P simply becomes

$$N_P = \left(\frac{1.2}{3.2}\right)^{3.3} = 0.03929$$

Thus:

$$N = \left[\frac{P}{N_P D_R^5 \rho_C}\right]^{1/3}$$

EXAMPLE 2.16.–

$$D_T = 2.59\ \text{m} \qquad \psi = 0.895\ \text{W.kg}^{-1}$$

$$D_R = 1.554\ \text{m} \qquad h = 0.2462\ \text{m}$$

$$P = \frac{\pi}{4} \times 0.895 \times 6.7081 \times 0.2462 \times 1100$$

$$P = 1277$$

$$N = \left[\frac{1277}{0.03929 \times 9.0627 \times 1100} \right]^{1/3} = 1.48 \text{ rev.s}^{-1}$$

We can verify that:

$$Re = \frac{1.48 \times 2.4149 \times 1100}{1.5.10^{-3}} = 26.2.10^{5} > 10^{5}$$

2.3.11. *Axial dispersion coefficients (dispersivities)*

Kumar and Hartland [KUM 92] give the expressions of the dispersion coefficients for the contiguous phase and the dispersed phase, which are D_{AC} and D_{AD}.

1) Continuous phase

$$D_{AC} = \frac{hV_C}{1-\phi}$$

$$\left[0.42 + 0.29\,R + A \times \left(\frac{V_C D_R \rho_C}{\mu_C} \right)^{-0.08} \times \left(\frac{D_T}{D_R} \right)^{0.16} \times \left(\frac{D_T}{h} \right)^{0.10} \times \left(\frac{D_S}{D_T} \right)^{2} \right]$$

where

$$A = 1.25.10^{-2} \left(\frac{ND_R}{V_C} \right) + \frac{13.38}{3.18 + ND_R / V_C}$$

2) Dispersed phase

$$\frac{D_{AD}}{V_D D_R} = 0.30 \left(1 + \frac{\overline{V}_C}{\overline{V}_D} \right) + 9.37 \left(\frac{ND_R}{\overline{V}_D} \right)^{-0.48} \times \phi^{-0.9} \times \left(\frac{D_R^2 \Delta\rho g}{\sigma} \right)^{-0.64} \times \left(\frac{D_T}{D_R} \right)^{0.7}$$

EXAMPLE 2.17.–

In light of the above results:

$$V_C = 0.223.10^{-3} m.s^{-1} \qquad \phi = 0.143 \qquad D_T = 2.59 \text{ m}$$

$$\bar{V}_C = \frac{0.223.10^{-3}}{1-0.143} = 0.260.10^{-3} m.s^{-1} \qquad N = 1.48 \text{ rev.s}^{-1}$$

$$h = 0.246 \text{ m} \qquad D_R = 1.554 \text{ m}$$

$$V_D = 0.29.10^{-3} m.s^{-1} \qquad R = 1.3 = \frac{V_D}{V_C}$$

$$\bar{V}_D = \frac{0.29.10^{-3}}{0.143} = 2.028.10^{-3} m.s^{-1}$$

$$hV_C/(1-\phi) = 0.246 \times 0.260.10^{-3} = 0.6396.10^{-4}$$

$$A = 1.26.10^{-2}\left(\frac{1.48 \times 1.554}{0.223.10^{-3}}\right) + \frac{13.38}{3.18 + 1.48 \times 1.554/0.223.10^{-5}}$$

$$A = 129.9513$$

$$D_{AC} = 0.6396\left[0.797 + 129.9513\left(\frac{0.223.10^{-3} \times 1.554 \times 1100}{1.5 \times 10^{-3}}\right)^{-0.08}\right] \times$$

$$0.6^{-0.16} \times \left(\frac{2.59}{0.246}\right)^{0.10} \times 0.7^2$$

$$D_{AC} = 0.003642 \text{ m}^2.s^{-1}$$

$$\frac{D_{AD}}{2.028.10^{-3} \times 1.554} = 0.30\left(1 + \frac{0.26}{2.028}\right) + 9.37\left(\frac{1.48 \times 1.554}{2.028.10^{-3}}\right)^{-0.48} \times$$

$$0.143^{-0.9} \times \left(\frac{1.554^2 \times 200 \times 9.81}{0.051}\right)^{-0.64} \times 0.6^{-0.7}$$

$$D_{AD} = 0.00107 \ m^2.s^{-1}$$

2.3.12. *Transfer coefficients*

We shall refer to the method put forward by Kumar and Hartland [KUM 99a, KUM 99b].

2.4. Pulsed column with perforated plates

2.4.1. *Structure of the device*

Horizontal plates do not have a spillway. Two consecutive plates define a compartment whose height is h:

$$5 \ \text{cm} < h < 10 \ \text{cm}$$

The plates contain holes whose diameter d is such that:

$$1 \ \text{mm} < d < 3 \ \text{mm}$$

The aperture of a plate is defined as the ratio of the surface of the holes to the total surface of the plate. We shall denote that aperture by the symbol "e":

$$0.2 < e < 0.25$$

If p is the step between the holes (center-to-center distance), we have:

– square arrangement:

$$e = \frac{\pi d^2}{4p^2}$$

– triangular arrangement:

$$e = \frac{\pi d^2}{2\sqrt{3}p^2}$$

The pulsations give a vertical to-and-fro motion to the liquid mass contained in the column. That motion is obtained by linking the base of the column to a volumetric pump (piston- or membrane-powered), containing no valves.

The amplitude A of the pulsations is the vertical distance between the extreme positions of the liquid content of the column.

$$1 \text{ cm} < A < 3 \text{ cm}$$

The frequency of the pulsations f is such that:

$$0.5 \text{ cycle.s}^{-1} < f < 2 \text{ cycle.s}^{-1}$$

There are no moving mechanical parts in the column because the pulsations are created by a pump, which may be some way away from the extractor.

2.4.2. *Operating regimes*

We can distinguish between three operating regimes, classified in *increasing* order of the product Af of the amplitude A and the frequency f of the pulsations:

1) mixer/settler: in each cycle, we distinguish four phases:

i) rest where, between two plates, there is settling with an interface separating two phases, i.e. separating the heavy species from the lighter one. There are no longer any drops,

ii) surge upwards: the lightweight species crosses the upper plate and, therefore, resolves into drops,

iii) rest and settling, as in phase 1,

iv) surge downwards: the heavy liquid crosses the lower plate and resolves into drops;

2) dispersion: the concentration in terms of number of drops is not uniform and is maximal:

i) near to the lower face of the plates, if the drops rise (the dispersed phase is light),

ii) near to the upper face of the plates, if the dispersed phase is the heavy phase;

3) emulsion: the concentration in terms of number of drops is uniform everywhere in the column. The diameter of the drops is of the order of 1–2 mm. Emulsion does not change the structure on crossing the plates and, in addition, during that journey, the drops continue to shift in relation to the contiguous phase. This operating regime is the most stable, and also that for which the diameter of the drops is smallest. Thus, we try, wherever possible, to work in the emulsion regime and, with this in mind, the product Af must be greater than or equal to a value of the order of $0.04\ m.s^{-1}$.

Experience shows us that, if we increase Af, we reduce the spreading of the distribution of the drop diameters. In addition, the flowrates of the two liquids have no influence on these diameters, so that the action of the product Af will suffice to obtain a precise value of the drop diameter.

As is the case in any extractor, we approach engorgement by excess flowrates when at least one of the two liquids is entrained out of the column toward the outlet pipe of the other liquid. The liquid thus entrained has not traversed the column. However, before the establishment of the flooding, an unstable operation may arise, during which the drops coexist with globules of dispersed phase, whose volume may be up to 100 times greater than that of a drop.

Note that the pulsed column with perforated plates presents an advantage over other extractors. Indeed, during a stoppage of around ten minutes, the dispersed phase remains in the compartments instead of propagating toward the outlet of that phase. This makes for a quick restart.

2.4.3. *Advantage of pulsed columns*

Pulsed columns, in comparison to packed columns:

– are capable of higher flowrates (for an equal diameter);

– have a higher transport efficiency: the HTUs are 2–3 times less;

– are invulnerable to interfacial fouling.

However, if the products have a natural tendency toward emulsification, or if the interfacial tension is low (less than 0.01 $N.m^{-1}$), the perforated fraction of the plates must be 40% or more.

Finally, this type of extractor must not be used if the liquids are sticky or greasy.

2.4.4. *Determination of the pulsation parameters*

According to Kumar and Hartland [KUM 98, KUM 99], the retention of the dispersed phase passes through a minimum when we vary the product Af of the amplitude A by the frequency of the cycles f. This minimum is at the medium of a transition zone between the (unacceptable) "mixer/settler" regime and the regime of "stable dispersion" which is to be sought.

For that minimum:

$$(Af)_m = 9.69.10^{-3}\left[\frac{\sigma\Delta\rho^{1/4}e}{\mu_d^{0.75}}\right]^{1/3} \qquad (m.s^{-1})$$

The retention increases exponentially from that minimum up until engorgement. We shall choose:

$$1.1(Af)_m < Af < 3(Af)_m$$

With the frequency f being chosen as between $0.5\,cycle.s^{-1}$ and $2\,cycle.s^{-1}$, we deduce the amplitude A.

σ : interfacial tension: $N.m^{-1}$;

$\Delta\rho$: difference between the densities of the 2 phases: $kg.m^{-3}$;

e : aperture of the plates (fraction of the which is holed);

μ_D : viscosity of the dispersed phase: Pa.s;

A : amplitude, which is the difference between the extreme heights achieved by the emulsion during the course of a cycle: m.

The aperture is taken as equal to 0.23 for the continuous phases whose viscosity is of the order of that of water, i.e. 0.001 Pa.s. It may reach as high as 0.4 if μ_D reaches a value of around 0.05 Pa.s.

EXAMPLE 2.18.–

$$\sigma = 0.015\ \text{N.m}^{-1} \qquad e = 0.23$$

$$\Delta\rho = 200\ \text{kg.m}^{-3} \qquad \mu_D = 2.10^{-3}\ \text{Pa.s}$$

$$(\text{Af})_m = 9.69.10^{-3}\left[\frac{0.015 \times 200^{1/4} \times 0.23}{\left(2.10^{-3}\right)^{0.75}}\right]^{1/3}$$

$$(\text{Af})_m = 0.01077\ \text{m.s}^{-1}$$

$$\text{Af} = 0.0118\ \text{m.s}^{-1}\ \#0.012\ \text{m.s}^{-1}$$

If we choose: $f = 1\ \text{cycle.s}^{-1}$

$$A = 0.012\ \text{m}$$

2.4.5. *Agitation power density*

According to Thornton *et al.*, [THO 57] and supposing that the vertical displacement of the mass of dispersion is a sine function of time, we can write:

$$\psi = \frac{\pi^2\left(1-e^2\right)}{2e^2\, C_o^2\, h}(\text{Af})^3 \qquad \text{(Watt per kg of dispersion)}$$

h : distance between two consecutive plates: m;

e : aperture of the plates (fraction of the surface which is holed);

C_0 : coefficient of contraction of the liquid vein for the holes.

EXAMPLE 2.19.–

$$e = 0.23 \qquad h = 0.05\text{ m}$$

$$C_0 = 0.61 \qquad Af = 0.012\text{ m.s}^{-1}$$

$$\psi = \frac{\pi^2\left(1-0.23^2\right)\times 0.012^3}{2\times 0.23^2\times 0.61^2\times 0.05}$$

$$\psi = 0.00821\ \text{W.kg}^{-1}$$

2.4.6. *Flooding*

We shall use the correlation by Smoot *et al.* [SMO 59]:

$$(V_C + V_D)_E = 0.527\left(\frac{\sigma}{\mu_C}\right)\left(\frac{V_D}{V_C}\right)^{0.014}\times\left(\frac{\Delta\rho}{\rho_C}\right)^{0.63}\times\left(\frac{\psi\mu_C^5}{\rho_C\sigma^4}\right)^{-0.207}\times$$

$$\left(\frac{d_0\sigma\rho_C}{\mu_C^2}\right)^{0.458}\times\left(\frac{g\mu_C^4}{\rho_C\sigma^3}\right)^{0.81}\times\left(\frac{\mu_C}{\mu_D}\right)^{0.20}$$

The approach to flooding – i.e. the ratio of the real flowrates to the flowrates on engorgement – is generally taken as between 0.5 and 0.7.

The diameter d_0 is that of the holes of the plates.

EXAMPLE 2.20.–

Besides the properties of the system under examination (which we have taken to be equal for all the extractors), certain parameters are specific to pulsed columns. Those which play a part here are:

$$d_0 = 0.003\text{ m} \qquad \text{and} \qquad \psi = 0.00821\ \text{W.kg}^{-1}$$

$$(V_C + V_D)_E = 0.527\left(\frac{0.015}{0.0015}\right)\times 1.3^{0.014}\times\left(\frac{200}{1100}\right)^{0.63}\times\left(\frac{0.00821\times 0.0015^5}{1100\times 0.015^4}\right)^{-0.207}\times$$

$$\left(\frac{3.10^{-3}\times 0.015\times 1100}{0.0015^2}\right)^{0.458}\times\left(\frac{9.81\times 0.0015^4}{1100\times 0.015^3}\right)\times\left(\frac{1.5}{2}\right)^{0.2}$$

$$(V_C + V_D)_E = 0.020\ \text{m.s}^{-1}$$

Let us choose an approach to flooding equal to 50%.

$$V_C = \frac{0.5\times 0.020}{1+1.3} = 0.00435\ \text{m.s}^{-1}$$

$$V_D = 1.3\times 0.00435 = 0.00565\ \text{m.s}^{-1}$$

2.4.7. *Mean drop diameter*

We shall use the correlation found by Kumar and Hartland [KUM 99]:

$$d_{32} = \frac{C_\psi e^{0.32} h}{\dfrac{1}{1.43\left(\dfrac{\sigma}{\Delta\rho g h^2}\right)^{1/2}} + \dfrac{1}{0.39\left[\dfrac{\psi}{g}\left(\dfrac{\Delta\rho}{g\sigma}\right)^{1/4}\right]^{-0.35}\left[h\left(\dfrac{\Delta\rho g}{\sigma}\right)^{1/2}\right]^{-1.15}}}$$

C_ψ is equal to 1 and 1.82 respectively for the transfers $C \rightarrow D$ and $D \rightarrow C$.

EXAMPLE 2.21.–

With the same data as before:

$$d_{32} = \frac{0.23^{0.32}\times 0.05}{\dfrac{1}{1.43\left(\dfrac{0.015}{200\times 9.81\times 0.05^2}\right)^{1/2}} + \dfrac{1}{0.39\left[\dfrac{0.00821}{9.81}\left(\dfrac{200}{9.81\times 0.015}\right)^{1/4}\right]^{-0.35}\left[0.05\left(\dfrac{200\times 9.81}{0.015}\right)^{1/2}\right]^{-1.15}}}$$

$$d_{32} = 1.30.10^{-3}\ \text{m}$$

The coefficient C_ψ was taken as equal to 1, which corresponds to a transfer from the continuous phase to the dispersed phase.

2.4.8. *Hold-up of the dispersed phase*

We shall use Kumar and Hartland's [KUM 99] expression.

$$\phi = \left[0.23 + \left\{\frac{\psi}{g}\left(\frac{\rho_C}{g\sigma}\right)^{1/4}\right\}^{0.71}\right]\left[V_D\left(\frac{\rho_C}{g\sigma}\right)^{1/4}\right]^{1.17} \exp\left[7.68V_C\left(\frac{\rho_C}{g\sigma}\right)^{1/4}\right]\times$$

$$\left[d_{32}\left(\frac{\rho_C g}{\sigma}\right)^{1/2}\right]^{-0.41}\left(\frac{\Delta\rho}{\rho_C}\right)^{-0.49}\left(\frac{\mu_D}{\mu_C}\right)^{0.82}\times 49.5\left[h\left(\frac{\rho_C g}{\sigma}\right)^{1/2}\right]^{-0.50}$$

EXAMPLE 2.22.–

With the above data:

$$\phi = \left[0.23 + \left\{\frac{0.00821}{9.81}\left(\frac{1100}{9.81\times 0.015}\right)^{1/4}\right\}^{0.71}\right]\left[0.00565\left(\frac{1100}{9.81\times 0.015}\right)^{1/4}\right]^{1.17}$$

$$\exp\left[7.68\times 0.00435\left(\frac{1100}{9.81\times 0.015}\right)^{1/4}\right]\times$$

$$\left[1.30.10^{-3}\left(\frac{1100\times 9.81}{0.015}\right)^{1/2}\right]^{-0.41}\left(\frac{200}{1100}\right)^{-0.49}\left(\frac{2.10^{-3}}{1.5.10^{-3}}\right)^{0.82}\times$$

$$49.5\left[0.05\left(\frac{1100\times 9.81}{0.015}\right)^{1/2}\right]^{-0.50}$$

$$\phi = 0.24$$

2.4.9. *Slip velocity*

This velocity is defined by:

$$V_{sl} = \frac{V_D}{\phi} + \frac{V_C}{1-\phi}$$

EXAMPLE 2.23.–

In view of the above results, we have:

$$V_{sl} = \frac{0.00565}{0.24} + \frac{0.00435}{0.76} = 0.02926 \text{ m.s}^{-1}$$

2.4.10. *Transfer coefficients*

We shall refer to the method proposed by Kumar and Hartland [KUM 99].

2.4.11. *Axial dispersion (contiguous phase)*

According to Kumar and Hartland [KUM 89], we can write:

$$\frac{D_{AC}}{\mu_C/\Delta\rho} = 46.15\exp(k\lambda)\left(\frac{V_D\mu_C}{\sigma}\right)^{0.11} \times \left(\frac{\mu_D}{\mu_C}\right)^{0.37} \times \left[\frac{(\sigma\Delta\rho h)^{1/2}}{\mu_C}\right]^{0.61} \times \left(\frac{d}{h}\right)^{0.36} \times \left(\frac{\Delta\rho h}{\rho_w \times 0.05}\right)^{1.05}$$

If $Af < 2(Af)_m$:

$$k = 0.80 \qquad \text{and} \qquad \lambda = \left[\frac{Af}{(Af)_m} - 1\right]^3 - \left[\frac{Af}{(Af)_m} - 1\right]^2$$

If $Af > 2(Af)_m$:

$$k = 0.34 \qquad \text{and} \qquad \lambda = \frac{Af}{(Af)_m} - 2$$

EXAMPLE 2.24.–

$$\mathrm{Af}/(\mathrm{Af})_m = 1.1 \qquad \text{so} \qquad \lambda = -0.009$$

$$\frac{D_{AC}}{7.5.10^{-6}} = 46.15\exp[0.8\times(-0.009)]\times\left(\frac{0.29.10^{-3}\times 1.5.10^{-3}}{0.015}\right)^{0.11}\times\left(\frac{2}{1.5}\right)^{0.37}\times$$

$$\left[\frac{(1.5.10^{-2}\times 200\times 0.05)^{1/2}}{1.5.10^{-3}}\right]^{0.61}\times\left(\frac{3.10^{-3}}{0.05}\right)^{0.36}\times\left(\frac{200}{1000}\right)^{1.05}$$

$$D_{AC} = 2.39.10^{-4}\ \mathrm{m^2.s^{-1}}$$

ρ_w is the density of water (1000 kg.m^{-3}).

2.4.12. *Axial dispersion (dispersed phase)*

According to Vermeulen *et al.* [VER 66], the coefficients corresponding to the two liquids can be considered equal.

3

Equilibrium and Material Transfer Between a Fluid and a Divided Solid

3.1. Introduction

In practice, an absorbent is characterized, in particular, by the following properties:

ρ_s: true density: $kg.m^{-3}$;

ε: porosity (fraction of void in volume);

a_m: surface per mass: $m^2.kg^{-1}$.

The distribution of the solute between fluid and solid will be obtained on the basis of the following concentrations:

q_V: quantity of solute absorbed per m^3 of solid;

c: quantity of solute dissolved per m^3 of fluid.

The surface per volume of the absorbent solid is:

$$a_V = a_m \times \rho_s \qquad \left(m^2.m^{-3}\right)$$

The quantity adsorbed per m^2 of solid surface is:

$$q_s = \frac{q_V}{a_V} \qquad \text{(kmol or kg per m}^2\text{)}$$

The quantity adsorbed per kg of absorbent is:

$$q_m = \frac{q_V}{\rho_S} \qquad \text{(kmol or kg per kg)}$$

In Chapter 3 and in [DUR 16b] we shall use only the concentrations c and q_V (which, for the sake of being concise, we shall simply call q).

To easily study the equilibria and transfers of material between a liquid and a divided solid, it would be ungainly to use a typical adsorption column. It is preferable to disperse a known mass of solid in a known volume of liquid. If the latter volume is of the order of magnitude of that of the solid, the variations in concentration in the liquid immediately give the variations in concentration q in the solid.

In order to study the couple gas–divided solid, we simply need to circulate a given mass of gas, in a closed loop, across the solid dispersed in a thin layer.

3.2. Choice of adsorbents

3.2.1. *General*

Remember that adsorption is the fixation of molecules of a fluid on the surface of a solid absorbent.

An adsorbent must be porous so as to allow the fluid to access the adsorption sites. The sites are distributed over the internal surface of the adsorbent, which may be as great as 1000 m^2 per gram of solid material. This internal surface is that of the walls of the pores whose diameter may be less than 2 nm.

The active porosity of the adsorbent corresponds to the volume of the nanopores, and may be up to 0.5. This porosity could be called the nanoporosity.

The macroporosity is that of the macropores (whose diameter is greater than 50 nm). The porosity of the macropores expresses the free volume situated between the agglomerates of nanoparticles associated with the nanopores. The macroporosity is useful to moderate the pressure drop in the fluid on crossing the adsorbent bed.

In Chapter 1 of [DUR 16c], the relationship between the diameter of the particles and the diameter of the pores is given.

Contact between the fluid and macropores takes place by convection. On the other hand, contact between the fluid and nanopores takes place by diffusion.

3.2.2. *Activated charcoal*

Activated charcoal is obtained from wood, lignin, tar, nut shells, artificial resins and heavy residues from oil refinement.

"Animal char", which is used to discolor sugar syrup, is obtained by pyrolysis of animal bones.

Activation takes place in two steps and in the absence of oxygen (H_2O and CO_2) in a Herreschoff kiln:

– at 450°C, the volatile components escape and therefore reveal nanopores;

– at 900°C is activation in the true sense.

The Herreschoff kiln includes superposed hearths, swept by rakes which ensure contact between the gas and the charcoal, but also encourages the material to progress toward the hearth immediately below the one in question, through a hole.

In general, activated charcoal is hydrophobic and readily adsorbs molecules with low polarity (phenol, benzene, toluene, organochlorines, etc.), whether in the gaseous or liquid phase. Detergents, pesticides and antibiotics are adsorbed in the liquid phase.

3.2.3. *Silica gel*

This gel has the chemical formula SiO_2 nH_2O and is polar. It therefore readily adsorbs water molecules. However, at low humidities, zeolite desiccation is preferable.

The adsorption of water by silica gel is an exothermic reaction (48×10^6 $J.kmol^{-1}$), unlike what happens with zeolites.

Silica gel regenerates at around 950°C in a stream of warm air. It is white, and is used in analytical, but also preparative, chromatography.

3.2.4. *Activated alumina*

This product, which has the formula $Al_2O_3\,nH_2O$ may be acidic or basic.

Activated alumina presents a great affinity for gaseous helium, hydrogen, argon, chlorine, HCl, SO_2, NH_3 and fluorocarbons. In the liquid phase, it is used for the drying of kerosene, aromatic compounds and essences.

3.2.5. *Activated earths*

Activated earths are clays, such as Fuller's earth (which is montmorillonite). They are used to discolor mineral oils, but today this operation is done by catalytic hydrogenation.

They readily adsorb colorants and toxic products. They are also used for the separation of isomers and even, it seems, for separating oxygen and nitrogen, though this separation actually tends to be achieved, in general, by cryogenic distillation of air.

3.2.6. *Ion exchangers*

The most stable exchangers are either slightly acidic ($5 < pH$) or slightly basic ($pH < 9$).

They are found:

1) in the form of gels. The porosity of the gel particles depends on their liquid content (swelling). The density of transverse connections lowers the kinetics of exchange and also decreases the stability;

2) in the form of porous particles obtained by numerous transverse connections. The diameter of the pores no longer varies with the liquid content.

Biomacromolecules whose diameter is approximately 10 nm need pore diameters of the order of 20–30 nm. For these molecules, it is preferable to use surfaces of hydrophilic resin. These resins are used in size-exclusion chromatography.

3.2.7. *Zeolites*

Zeolites are mixtures of SiO_4 and AlO_4 groups. Today, they are obtained by synthesis.

They form sorts of cages to which access is gained through apertures of a very specific size, and for this reason they are known as molecular sieves. The diameter of these apertures, depending on the type of zeolite, may be equal to 0.3, 0.4, 0.5 or 1 nanometer. Consequently, the active surface concept does not apply to zeolites.

If the silica/alumina ratio increases, the zeolite becomes hydrophobic. These products are non-combustible and can therefore be regenerated by hot air.

At low humidities, molecular sieves of 0.3 nm in aperture work very well.

3.3. The different types of isotherms of fluid–solid equilibrium

3.3.1. *The Freundlich isotherm*

This isotherm was first used by Boedecker [BOE 85]. It is written:

$$q = ac^{1/n}$$

This is a purely empirical equation.

q : concentration of the solute in the solid: $kmol.kg^{-1}$ or $kg.kg^{-1}$;

c : concentration of the solute in the liquid: $kmol.m^{-3}$ or $kg.m^{-3}$.

The so-called Freundlich isotherm may be used for gases. In this case, we write:

$$q = ap^{1/n}$$

p : partial pressure of the component in the gaseous phase: Pa.

3.3.2. *Langmuir–Freundlich combination*

This isotherm is usable for the adsorption of organic solutes onto carbon.

According to Fritz and Schluender [FRI 74] and to Umpleby *et al.* [UMP 01], it is written:

$$q_i = \frac{a_{oi} c_i^{b_{oi}}}{1 + \sum_{i=1}^{n} a_{si} c^{b_{si}}}$$

3.3.3. *The Langmuir isotherm (thermodynamic approach [VOL 25]*

If we take account of the surface β of an absorbed molecule, we can modify the ideal gas equation and obtain:

$$\pi(A - \beta) = n_{ads} RT$$

Hence, if we accept that β^2 is negligible:

$$\frac{\partial \pi}{\partial A} = -\frac{n_{ads} RT}{(A-\beta)^2} \# -\frac{n_{ads} RT}{A^2 - 2\beta A} \qquad [3.1]$$

The Gibbs isotherm is (ideal gas law):

$$\frac{A d\pi}{n_{ads} RT} - A d\pi = n_{ads} d\mu_{ads} = n_{ads} RT \frac{dP}{P} \qquad [3.2]$$

Let us eliminate between equations [3.1] and [3.2]:

$$\frac{dP}{P} = -\frac{dA}{A - 2\beta}$$

By integrating, we find:

$$P = \frac{K}{A - 2\beta}$$

Let $\theta = 2\beta/A$ be the degree of coverage of the surface A. We find:

$$P(1-\theta)=\frac{K}{A} \quad [3.3a]$$

Let us posit:

$$\frac{Kb}{A}=\frac{q}{q_s}=\theta$$

Equation [3.3] becomes the Langmuir isotherm:

$$\theta=\frac{bP}{1+bP} \quad [3.3b]$$

3.3.4. *The Langmuir isotherm (dynamic equilibrium) [LAN 16, LAN 18]*

When a fluid is at equilibrium with a solid, there is equilibrium between:

$$\alpha\theta_0\mu \quad = \quad \nu_1\theta_1 \quad [3.4]$$

condensation evaporation

θ_0 : fraction of the surface without adsorbate;

θ_1 : fraction of the surface with adsorbate (coverage rate);

μ : flowrate of molecules striking the surface: molecule.s^{-1}.m^{-2};

α : fraction of incident molecules which remain fixed to the wall;

ν_1 : flowrate of the evaporated molecules: molecule.s^{-1}.m^{-2}.

Naturally:

$$\theta_1+\theta_0=1 \quad [3.5]$$

By setting:

$$\frac{\alpha}{\nu_1}=\sigma$$

and by eliminating θ_0 between equations [3.4] and [3.5]:

$$\theta_1 = \frac{\sigma\mu}{1+\sigma\mu}$$

with, for a gas:

$$\mu = \left(\frac{1}{2\pi RTM}\right)^{1/2} \times P$$

P : gas pressure: Pa;

R : ideal gas constant: 8314 $J.kmol^{-1}K^{-1}$;

T : absolute temperature: K;

M : molar mass of the gas: $kg.kmol^{-1}$.

This justification was put forward by Langmuir [LAN 18].

The isotherm can then be written:

$$\theta = \frac{bP}{1+bP} \text{ where } b = \sigma\mu \qquad [3.6]$$

3.3.5. *Generalization to multiple components*

We shall suppose that σ has the same value for all the components, and that only the coefficient μ_i is particular to each component with index i.

We set:

$$\mu = \sum_{i=1}^{n} \mu_i y_i \text{ and } b = \sigma \sum_{i=1}^{n} \mu_i y_i = \sum_{i=1}^{n} b_i y_i$$

Equation [3.3b] becomes:

$$\theta = \frac{\sum_{i=1}^{n} b_i y_i P}{1+\sum_{i} b_i y_i P} = \frac{\sum_{1}^{n} b_i p_i}{1+\sum_{i} b_i p_i} \qquad \left(\text{when } \theta = \sum_{i-1}^{n} \theta_i\right)$$

This relation is compatible with:

$$\theta_i = \frac{b_i p_i}{1+\sum_{i=1}^{n} b_i p_i}$$

3.3.6. *Generalization to a heterogeneous surface (bi-Langmuir isotherm)*

Let h_1 and h_2 represent the fractions of the surface occupied respectively by sites 1 and 2. We have:

$$h_1 + h_2 = 1$$

We can write an isotherm separately for each type of site:

$$\theta = \theta_1 + \theta_2 = \frac{h_1 b_1 P}{1+b_1 P} + \frac{h_2 b_2 P}{1+b_2 P}$$

3.3.7. *Langmuir equation for liquids*

We merely need to replace the partial pressure of the gas with the activity ("a") of the solute in the solution:

$$\theta = \frac{ba}{1+ba}$$

Indeed, in solutions, the activity coefficient of the solute is rarely equal to 1.

3.3.8. *"Competitive" isotherm found by Gritti and Guiochon [GRI 03]*

The Langmuir equation for different components (in competition for adsorption) does not satisfy the Gibbs–Duhem equation (see section 1.8.1 in [DUR 16a]).

For this reason, Gritti and Guiochon [GRI 03] proposed a "competitive" isotherm (i.e. for a mixture) which, for its part, does satisfy the Gibbs–

Duhem equation or, more specifically, the equality of the spreading pressures.

The parameters of this isotherm are obtained on the basis of the parameters of the isotherms of each component taken in isolation.

3.3.9. *Tóth isotherm [TÓT 71]*

Tóth proposed the following equation:

$$n_s = \frac{n_{s\infty} p}{\left(b + p^m\right)^{1/m}}$$

The parameter m characterizes the homogeneity of the absorbent:

m = 1 – completely homogeneous surface;

m = 0 – completely heterogeneous surface.

However, experience shows that m may vary depending on the nature of the adsorbate, which takes away considerably from the advantage to using Tóth's equation.

3.3.10. *Moreau isotherm [MOR 91]*

The isotherm is presented in the following form:

$$q = q\infty \frac{bc + Ib^2c^2}{1 + 2bc + Ib^2c^2}$$

q_s: capacity at saturation;

b: equilibrium constant at a concentration close to zero;

I: parameter of interaction between two identical molecules;

There is a so-called bi-Moreau isotherm.

3.3.11. *Martire isotherm [MAR 87, MAR 88]*

This isotherm involves numerous parameters. A limiting case is the Langmuir isotherm. Martire's isotherm is of use for heterogeneous surfaces.

3.3.12. *BET (Brunauer, Emmett and Teller) isotherm [BRU 38]*

Brunauer, Emmett and Teller based their reasoning on the fact that:

– adsorption takes place in multiple superposed layers;

– the first layer is linked to the solid by a veritable energy of adsorption;

– for the other layers, each is linked to the previous one by the energy of liquefaction of the adsorbate in the pure state.

The isotherm is written as follows for the gases:

$$q = q_s \frac{k\,p_r}{(1-p_r)\left[(k-1)p_r + 1\right]}$$

q_s : charge at saturation: $kmol.kg^{-1}$

$$p_r = P_v / \pi \text{ (relative pressure)}$$

P_v : partial pressure of the gaseous adsorbate: Pa;

$\pi(t)$: vapor pressure of the adsorbate: Pa.

Rounsley [ROU 60], and later, El Sabawi and Pei [EL 77], supplemented the BET equation to make it exact throughout the range from zero to one for the relative humidity of the gaseous phase. These authors drew a distinction between, firstly, multilayer adsorption to the wall of empty pores and, secondly, the gradual filling of those pores by the liquid.

The BET equation can also be used when the adsorbate comes from a liquid phase. We then write:

$$q = q_s \frac{b_0 c}{(1-b_l c)(1+b_0 c - b_l c)}$$

c: concentration in the liquid phase: $kmol.m^{-3}$;

b_0: constant for adsorption to the naked surface;

b_ℓ: constant for adsorption to the previous layer.

3.3.13. *Ion exchange*

Consider a resin in the divided state, linked to an ion A whose valence is ν_A. Also consider an ion B linked to an ion with the opposite charge X. The valence of B is ν_B. The ion exchange is written as follows:

The law of mass action is written:

$$K_0 = \frac{c_B^{\nu_A} q_A^{\nu_{B0}}}{q_B^{\nu_A} c_A^{\nu_B}}$$

The activity coefficients are supposed to be incorporated into the constant K.

More specifically, suppose that we have:

$$\nu_A = \nu_B = \nu$$

Let us divide the top and bottom by $c_0^{\nu} = (c_A + c_B)^{\nu}$ and by $Q = (q_A + q_B)^{\nu}$:

$$\frac{c_A}{c_0} = x \qquad \frac{c_B}{c_0} = 1 - x \qquad \frac{q_A}{Q} = y \quad \text{and} \quad \frac{q_B}{Q} = 1 - y$$

We obtain:

$$K = K_0^{1/\mu} = \frac{y(1-x)}{x(1-y)} = \frac{q_A}{c_A} \times \frac{c_B}{q_B}$$

Thus:

$$y = \frac{Kx}{1 + (K-1)x}$$

If K > 1, the isotherm is said to be langmuirian.

If K < 1, the isotherm is said to be antilangmuirian.

When, for example, μ_A=2 and μ_B=1, the fraction y is obtained by solving a second-degree equation.

NOTE.–

Let us more closely examine the equivalence between the Langmuir law and an ion-exchange equilibrium.

The Langmuir law is written:

$$q = \frac{q_\infty Ac}{1 + Ac}$$

If c increases indefinitely, then q tends toward q_∞.

In addition, for an exchange of ions of the same valence, the equilibrium is written:

$$c(Q-q)K = q(c^* - c)$$

Q: maximum capacity of adsorption of the solid.

In the equilibrium equation, we see that if $q = Q$, then necessarily, $c = c^*$ as well. In other words, it is the concentration at equilibrium with Q.

The equilibrium equation can be written:

$$q = \frac{QKc}{c^* + (K-1)c} = \frac{QKx}{1 + (K-1)x} \quad \text{where} \quad \frac{c}{c^*} = x$$

For its part, the Langmuir law can be written:

$$q = \frac{q_\infty Bx}{1 + Bx} \quad \text{when} \quad \frac{c}{c^*} = x \quad \text{and} \quad \frac{B}{c^*} = A$$

The equivalence of the two expressions of q as a function of x means that:

$$B = K - 1 \text{ and } Q = \frac{q_\infty (K-1)}{K}$$

and also:

$$K = 1 + B \text{ and } q_\infty = \frac{Q(1+B)}{B}$$

3.3.14. *Generalization*

Gas–solid $\qquad P_i = y_i P = z_i P_i^*(\pi)$

Liquid–solid $\qquad c_i = x_i c_T = z_i c_i^*(\pi)$

Consequently, if we write that $\sum_i z_i = 1$

$$\sum_i \frac{P_{Vi}}{P^*(\pi)} = 1 \text{ and } \sum_i \frac{c_i}{c_i^*(\pi)} = 1 \qquad [3.7]$$

$P_i^*(\pi)$ and $c_i^*(\pi)$ can be inverted, because they are monotonic increasing functions, and we can write:

$$\pi_i = \pi_i(P_i^*) \quad \text{and} \quad \pi_i = \pi_i(c_i^*)$$

By writing (n-1) equations between the values π_i and taking account of equation [3.7], we obtain n equations which can be used to calculate the $n\ P_i^*$ or the $n\ c_i^*$.

Then, having taken a value for the P_i or the c_i, we calculate:

$$z_i = \frac{P_i}{P_i^*} \text{ or indeed } z_i = \frac{c_i}{c_i^*}$$

Brunauer *et al.* [BRU 40] put forward a rather complicated equation known as the "E equation" which, in their view, should be able to account for the five known main forms of isotherms. However, Coulson *et al.* [COU 79] proposed a simpler equation which, according to those authors, is also able to express the five types of isotherms.

$$q_s^* = q_1 By \frac{1-(n+1)y^n + ny^{n+1}}{(1-y)\left[1+(B-1)y - By^{n+1}\right]} \quad \text{where } y = \frac{P_i}{P} \text{ or } \frac{c_i}{c_T}$$

When n is greater than or equal to 3, we can, by choosing an appropriate value for B, obtain an isotherm of the desired form. With this isotherm, we can calculate:

$$\pi = \frac{RT}{MS}\int_0^{P_i^*} q\left(P_i^*\right)\frac{dP_i}{P_i} = \pi\left(P_i^*\right) \text{ or indeed } \pi\left(c_i^*\right)$$

It is always possible to use other analytical forms of isotherms, depending on the problem at hand, which may simplify the calculations.

3.3.15. *The five main types of isotherms*

It is possible to assign an analytical expression to each type of isotherm. These isotherms are of the form $q = f(P_V/\pi(t))$ and are therefore appropriate for gases:

1) Freundlich (see section 3.3.1);

2) BET (see section 3.3.12);

3) anti-langmuirian;

4) Rounsley [ROU 60] and also El Sabawi and Pei [EL 77] (see Volume 4).

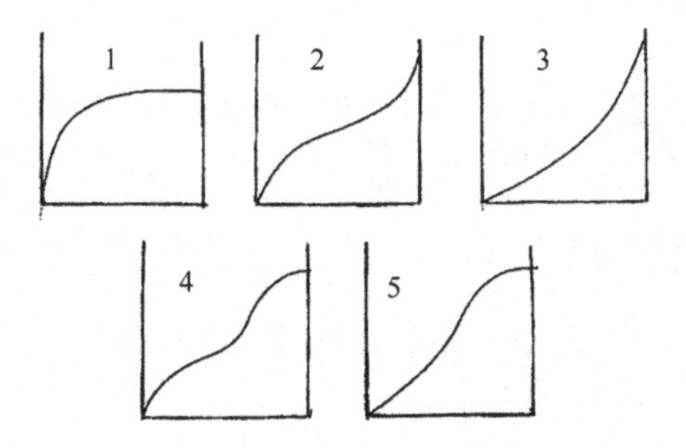

Figure 3.1. *Types of isotherms q as a function of P_V/P_T or c/c_∞*

Type-4 isotherms, for high values of q and P_V, contain a part which conforms to the laws of capillarity (imbibition and drainage, see Chapter 4). In that part, a hysteresis loop is highly possible.

Types 1 and 2 can be adapted for liquids.

3.4. Thermodynamics and equilibrium of multiple adsorption

3.4.1. *Posing of the problem*

We take a mass m of a divided solid and bring that mass into contact with a fluid (liquid or gas) characterized by its concentration of different solutes.

It is understood that the mass of the solid is small in comparison to that of the fluid, so the adsorption of the solutes to the solid has no noticeable effect on the content of the fluid.

Therefore, it is a question of determining:

– composition in molar fractions z_i of the adsorbate

– total number of kmoles adsorbed.

The discussion below is based on the publications of Myen and Prausnitz [MYE 65], and Radke and Prausnitz [RAD 72].

3.4.2. *Spreading pressure – Gibbs equation*

Gibbs made the hypothesis of the presence, between the fluid and the solid, of an interface with thickness 0, where a number n_s of absorbed molecules are to be found.

The surface energy of those molecules has decreased ($\Delta\gamma < 0$). For the elementary surface dA, we posit:

$$-\Delta\gamma dn_s = \pi dA > 0$$

By definition, π is the positive spreading pressure. *This is the pressure exerted by liquid on the solid.*

The Helmholtz energy of the absorbed molecules is, at constant temperature:

$$dF_s = \mu_s \, dn_s - \pi dA \qquad [3.8]$$

Note that the term –PdV involving the volume has disappeared, because the thickness of the interface is null. It has been replaced by the term $-\pi dA$.

Let us integrate with constant intensive variables. We obtain the Helmholtz energy of the adsorbed phase:

$$F_s = \mu_s n_s - \pi A \quad [3.9]$$

Let us now take the total differential:

$$dF_s = n_s \, d\mu_s + \mu_s dn_s - \pi dA - A d\pi \quad [3.10]$$

We now subtract equations [3.8] and [3.10] from one another, term by term. We obtain the Gibbs equation:

$$n_s \, d\mu_s - A d\pi = 0$$

We set:

$$\Gamma_s = \frac{n_s}{A} \text{ and } f_s = \frac{F_s}{A}$$

Equation [3.9] becomes:

$$f_s = \Gamma_s \mu_s - \pi$$

It is easy to generalize the above results to multiple absorbed species, by setting:

$$f_s = \sum_i \Gamma_{si} \mu_{si} - \pi$$

NOTE.–

Lennard-Jones [LEN 32] studied the displacement of atoms or molecules absorbed to the surface of the absorbent solid. He also studied absorption into the solid from that surface.

3.4.3. *Surface energy and surface tension of liquids [LAN 16]*

Consider the interface separating two immiscible phases. That interface is characterized by:

– a surface A;

– an entropy S.

Let A and S be modified by an external action. The variation in surface energy of the interface is defined by:

$$dU = \gamma dA + TdS \quad [3.11]$$

The resulting variation in Helmholtz energy is:

$$dF = d(U - TS) = \gamma dA - SdT$$

Therefore:

$$\frac{\partial F}{\partial A} = \gamma \text{ and } \frac{\partial F}{\partial T} = -S \text{ and } \frac{\partial^2 F}{\partial A \partial T} = \frac{\partial \gamma}{\partial T} = \frac{-\partial S}{\partial A}$$

The variation in energy expressed per unit surface is, in light of equation [3.11]:

$$\frac{\partial U}{\partial A} = \gamma + T\frac{\partial S}{\partial A} = \gamma - T\left(\frac{\partial \gamma}{\partial T}\right)$$

All of this assumes that:

– the interfacial Helmholtz energy is γ A;

– what we call the total interfacial energy is U;

– the surface tension is γ.

In general, $\partial\gamma/\partial T$ is negative. Therefore, the total surface energy is positive (γ is positive by definition).

3.4.4. *Surface tension and spreading pressure*

We have written the surface term in two different ways in the expression of the Helmholtz energy.

$$F = \gamma A + \text{etc.}\dots \text{ and } F = -\pi A + \text{etc.}\dots$$

This should come as no surprise. Indeed, as happens in divided solid mechanics, a tension and a compression are negative and positive, respectively. Strictly speaking, therefore, we ought to write:

$$F = -\gamma A + \text{etc.}\dots \text{ where } \gamma < 0$$

3.4.5. *Gibbs equation and tensioactives*

Tensioactives in solution in a liquid accumulate at the interface separating the solution from one of the following three phases:

– another liquid which is immiscible with the first;

– a gaseous phase;

– or a solid phase.

For example, a tensioactive compound would accumulate at the interface between an aqueous phase and a hydrophobic phase.

The concentration of tensioactive species at the interface, therefore, is higher than it is in the solution. In order to determine the number of kmoles of tensioactives at the interface, we simply need to find the difference between the initial concentration of the liquid and its concentration once migration toward the interface has taken place.

In general, we use this difference n_t expressed in relation to the area A of the interface.

$$\Gamma_s = \frac{n_t}{A}$$

The Gibbs equation then becomes:

$$\Gamma_s d\mu_s - d\pi = 0$$

3.4.6. *Calculation of the spreading pressure on a solid*

According to the Gibbs equation:

$$d\pi = \frac{n_s}{A} d\mu_s = \frac{q}{MS} d\mu_s \qquad [3.12]$$

q: mass of adsorbate per kg of absorbent: $kg.kg^{-1}$;

M: molar mass of adsorbate: $kg.kmol^{-1}$;

S: surface occupied by a kmole of adsorbate $m^2.kmol^{-1}$.

The mass q depends on the partial pressure of the adsorbate P_V, if the fluid is a gas, and the concentration c_s if the fluid is a liquid.

The chemical potential of the adsorbate is, with the difference of a constant:

RT Ln P_V (for a gas) and RT Ln c (for a liquid)

Equation [3.12] can then be integrated, for example, for a gas and thus give us the spreading pressure *for a single adsorbate.*

$$\pi = \frac{RT}{MS}\int_0^{P_S} q(P_V)\frac{dP_V}{P_V}$$

P_V: partial pressure of the adsorbate in the gas: Pa;

S: surface of the absorbent per unit mass of absorbent: m^2kg^{-1}.

The integral converges at its lower bound if there is a Henry constant.

$$q(P_V) = HP_V \text{ or } q(c) = Hc$$

3.4.7. *Padé approximation*

Frey and Rodrigues [FRE 94] use this approximation to express the vapor pressure of a component i as a function of the spreading pressure π.

$$P_i^* = \frac{\pi}{a_i + f_i\pi + c_i\pi^2}$$

The advantage of this expression is that it can easily be inverted following the resolution of a second-degree equation.

3.4.8. *Equilibrium equations; composition of the adsorbate*

The expression of the spreading pressure is, for the component i

$$\pi_i = f_i\left(c_s^*\right) \text{ or } \pi = f_i\left(P_s^*\right)$$

The spreading pressure must be common to all the absorbed species.

$$\pi_i = \pi$$

Let z_i be the molar fraction of the species i in the Gibbs film.

$$\sum_{i=1}^{n} z_i = 1$$

If the fluid phase contains a component that is not absorbed, we still need to take account of it with a very low value for that component in the Gibbs film.

Thus, we have determined the c_i^* or the P_i^*.

The n equilibrium equations are therefore written:

$$y_i P = z_i P_i^* \text{ or } y_i c_\tau = z_i c_i^*$$

Naturally, it is always possible to replace the molar fractions z with activities $a_i = z_i\gamma_i$. The γ_i might, for instance, be calculated using Wilson's formulas (see section 3.1.5 in [DUR 16a]). Often, though, researchers accept the hypothesis that the γ_i are equal to 1.

Finally, we have (n-1) equations if dealing with a gas.

$$\int_0^{\frac{y_1 P}{z_1}} \frac{q_1}{M_1 S} \frac{dP_{V1}}{P_{V1}} = \int_0^{\frac{y_2 P}{z_2}} \frac{q_2}{M_2 S} \frac{dP_{V2}}{P_{V2}} = \ldots = \int_0^{\frac{y_n P}{z_n}} \frac{q_n}{M_n} \frac{dP_{Vn}}{P_{Vn}}$$

If we add to these (n-1) equations the fact that the sum of the z_i is equal to 1, it is possible to calculate the n z_i.

3.4.9. *Calculating the molar sum of the adsorbed species*

The fugacity of the adsorbed compound is written as follows:

$$f_{si} = z_i f_{si}^*$$

The chemical potential of the adsorbed i is:

$$\mu_{si} = RT \ln f_{si} = RT \ln z_i + RT \ln f_{si}^*$$

Thus:

$$\left.\frac{d\mu_{si}}{d\pi}\right|_{z_i=\text{const.}} = \frac{d\mu_{si}^*}{d\pi}$$

The asterisk refers to the adsorption of the compound i from its pure fluid form. In this case, the Gibbs equation is written:

$$\frac{A}{n_{si}^*} = \frac{d\mu_{si}^*}{d\pi}$$

For the total mixture, the Gibbs equation is written:

$$Ad\pi = \sum_i n_{si}^* d\mu_{si} \qquad Ad\pi = \sum_i n_{si}^* d\mu_{si}^*$$

Hence, after division by $d\pi$ and by the sum of the n_T absorbed molecules:

$$\frac{A}{n_T} = \sum_i z_i \frac{d\mu_{si}}{d\pi} = A\sum_i z_i \frac{d\mu_{si}^*}{d\pi} = \sum_i A\frac{z_i}{n_{si}^*}$$

and finally:

$$\frac{1}{n_T} = \sum \frac{z_i}{n_{si}^*}$$

3.4.10. *Comparison with the Raoult laws*

These laws, in the case of ideality, are expressed by:

$$y_i P = x_i \pi_i (T)$$

$\pi_i(T)$: saturating vapor pressure in the pure state of the compound i: Pa.

By bringing into play what could be called the saturating vapor pressure of the adsorbed phase, we can write:

$$y_i P = z_i P_{si}^* (\pi)$$

In this last formula, π is the spreading pressure.

In Raoult's laws, the temperature is the same for all the components, whilst in the adsorption laws, the spreading pressure must be the same for all the components (as, of course, must the temperature also). This explains the interest held by the work of Frey and Rodrigues [FRE 94].

Lewis *et al.* [LEW 50] had already shown that between two adsorbates, it is possible to introduce a constant relative volatility. Grant and Manes [GRA 66] did likewise:

$$\alpha = \frac{y_1 z_2}{z_1 y_2} = \frac{P_{s1}(\pi)}{P_{s2}(\pi)}$$

Myers [MYE 83] proposed activity coefficients different to 1 for non-ideal mixtures.

3.4.11. *Fick's laws and activity coefficient:*

Fick's law is written:

$$N = -\frac{D}{RT} q \frac{d\mu}{dZ}$$

q: concentration of the adsorbate *in* the solid: kg.m^{-3} or kg.m^{-3};

Z: dimension in the direction of the flux: m.

The chemical potential $\mu(q)$ must be expressed directly as a function of q, rather than as a function of a particular relative pressure or concentration of the phase which would be at equilibrium with the solid [GAR 72].

Similarly to liquid–vapor equilibrium and liquid–liquid equilibrium, the chemical potential of the adsorbate can be written:

$$\mu = \mu_0 + RTLn\gamma_s q$$

γ_s : activity coefficient.

Fick's law is then written:

$$N = -D\left[1 + \frac{dLn\gamma_s}{dLnq}\right]\frac{dq}{dZ}$$

Whilst experience shows that Fick's law may be written:

$$N = -Df(q)\frac{dq}{dZ}$$

We simply need to integrate the following relation to find the expression of γ:

$$dLnq + dLn\gamma = f(q)dLnq$$

3.4.12. *Calculation of γ_s with a linear isotherm*

The linear isotherm is expressed by:

$$q = ac \quad ; \quad \frac{dq}{dc} = a$$

Hence:

$$\frac{d\mu}{dz} = \frac{d\mu}{dc} \times \frac{dc}{dq} \times \frac{dq}{dz} = \frac{RT}{c} \times \frac{1}{a} \times \frac{dq}{dz} = RT\frac{dLnq}{dz}$$

This latest equality shows that $\gamma_s = 1$, and *this characterizes numerous calculations of diffusion in a solid.*

3.4.13. *Calculation of γ_s with the Langmuir isotherm*

$$\frac{d\mu}{dz} = \frac{d\mu}{dq} \times \frac{dq}{dz} = \frac{d\mu}{dc} \times \frac{dc}{dq} \times \frac{dq}{dz}$$

$\mu = RTLnc$ (we suppose that $\gamma_s = 1$ in the liquid).

In light of the Langmuir isotherm:

$$q = \frac{q_\infty ac}{1+ac}$$

Thus:

$$c = \frac{q}{a\left(q_\infty - q\right)} = \frac{qs}{a\left(q_\infty - q\right)} - \frac{1}{a}$$

$$\frac{dc}{dq} = \frac{q_\infty}{a\left(q_s - q\right)^2} \qquad \frac{d\mu}{dc} = \frac{RT}{c} = \frac{RTa\left(q_\infty - q\right)}{q}$$

According to (1):

$$\frac{d\mu}{dq} = \frac{q_\infty}{a\left(q_\infty - q\right)} \times \frac{RTa\left(q_\infty - q\right)}{q}$$

Hence:

$$q\frac{d\mu}{dq} = -\frac{RTq_\infty}{\left(q_\infty - q\right)}$$

$$N = -\frac{D}{RT}q\frac{d\mu}{dz} = -\frac{Dq_\infty}{q_\infty - q}\frac{dq}{dz}$$

However, as we know from section 3.4.11:

$$\frac{q_\infty}{q_\infty - q} = 1 + \frac{dLn\gamma_s}{dLnq}$$

$$dLn\gamma_s = \left(\frac{q_\infty}{q_\infty - q} - 1\right)\frac{dq}{q}$$

$$Ln\gamma_S = -\int_0^q dLn(q_\infty - q)dq$$

$$\gamma_s = \frac{q_\infty}{q_\infty - q}$$

and:

$$\mu_s = \mu_s^0 + RT\,Ln\left(\frac{qq_\infty}{q_\infty - q}\right)$$

We can see that γ_s tends toward 1 as q tends toward zero.

We also see that a (widespread) mistake is made in writing Fick's law for the Langmuir isotherm:

$$N = -D\frac{dq}{dz}$$

whereas in actual fact, we need to write:

$$N = -Dq\frac{d\mu}{dz} = -\frac{Dq_\infty}{q_\infty - q}\frac{dq}{dz}$$

The calculations remain simple if q is negligible in relation to q_∞, but if not, we can see that certain results in the existing literature need to be revised.

The expression of γ_s thus obtained depends on the analytical form of the isotherm chosen. We could perform the same calculation with the BET isotherm or with Coulson and Richardson's [COU 79], whose physical direction is more accurate than the Langmuir isotherm. Thus, Gritti and Guiochon [GRI 03] chose the BET isotherm to express the simultaneous equilibrium of two solutes.

3.4.14. *Isosteric heat [MIY 01]*

Consider an adsorption. The chemical potentials in the fluid and the solid are μ_{fl} and μ_{sol}.

We accept that the activity coefficients are equal to 1.

$$\Delta G = \mu_{sol} - \mu_{fl} = \mu_{sol}^0 + RT_0 Lnq - \mu_{fl}^0 - RT_0 Lnc$$

$$\Delta G = \Delta G_0 + RT_0 LnK$$

where

$$\mu_{sol}^{0} - \mu_{fl}^{0} = \Delta G_0 \quad \text{and} \quad K = \frac{q}{c}$$

In general, adsorptions are exothermic.

$$\Delta Q < 0$$

It stems from this that the isosteric heat is positive because, by definition:

$$Q_{ist} = -\Delta Q$$

When the temperature varies from T_0 to T, the entropy of the system varies by

$$\Delta S = \Delta Q\left(\frac{1}{T} - \frac{1}{T_0}\right) = -Q_{ist}\left(\frac{1}{T} - \frac{1}{T_0}\right)$$

The authors write that $\Delta H = \Delta G + T_0\Delta S = 0$

$$\Delta G = \Delta G_0 + RT_0 \text{LnK} = -T_0\Delta S$$

or indeed:

$$\text{LnK} = \frac{Q_{ist}}{R}\left(\frac{1}{T} - \frac{1}{T_0}\right) - \frac{\Delta G_0}{RT_0}$$

The slope and the ordinate at the origin of $\text{LnK} = f(Q_{ist})$ give us T_0 and ΔG_{0T}.

To achieve this result we must, at a constant temperature, perform adsorptions using the same solvent but varying the nature of the adsorbate. Indeed, each adsorbate has a corresponding value of Q_{ist}.

Naturally, for a give solvent, adsorbate and adsorbent, LnK is a linear function of 1/T.

3.5. Transfer parameters

3.5.1. *Transfer through the liquid film*

Each particle is surrounded by a liquid film, through which the solute must pass in one direction or the other.

An intense agitation activates this crossing by decreasing the thickness of the film. See section 2.3.2 in [DUR 16d] for the calculation of the transfer flux density through the liquid film.

3.5.2. *Expression of the Knudsen diffusivity in a pore*

Two previous results are needed:

– mean quadratic velocity of the gas [DYE 66]

$$\overline{v} = \left(\frac{8kT}{\pi m} \right)^{1/2}$$

– mean free path of a gas [GIL 58]

$$\overline{\lambda} = \frac{3\mu}{P} \left(\frac{\pi kT}{8m} \right)^{1/2}$$

k : Boltzmann's constant: 1.38048×10^{-23} $J.K^{-1}$;

m : molar mass of the gas: $kg.kmol^{-1}$;

T : absolute temperature: K;

μ : viscosity of the gas: Pa.s;

P : gas pressure: Pa;

$\overline{\lambda}$: mean free path: m;

$\overline{v}$: mean quadratic velocity: $m.s^{-1}$.

The Knudsen diffusivity D_k comes into play *when the diameter d_p of the pores is less than or equal to the mean free path of the gas.*

$$d_p \leq \frac{3\mu}{P}\left(\frac{\pi kT}{8m}\right)^{1/2}$$

$$D_K = \frac{\overline{v}}{3} d_p = \frac{d_p}{3}\left(\frac{8kT}{\pi m}\right)^{1/2}$$

NOTE.–

The rarefied gas-flow theory shows that the viscosity of such a gas is:

$$\mu = \frac{m\overline{v}}{2^{3/2}\pi d_m^2}$$

d_m : molecular diameter: m;

μ : viscosity: Pa.s.

EXAMPLE.–

$D_{pore} = 10^{-9}$ m $\quad M_{CO_2} = 44$ kg.kmol^{-1} $\quad N_A = 6.023 \times 10^{26}$ molecule.kmole^{-1}

$P = 10^5$ Pa $\quad d_m = 0.32$ nm $\quad T = 293$ K

$$m = \frac{44}{6.023.10^{26}} = 7.305.10^{-26}.\text{kg}$$

$$\overline{v} = \left(\frac{8 \times 1.38048.10^{-23} \times 293}{\pi \times 7.305.10^{-26}}\right)^{1/2}$$

$$\overline{v} = 375\text{m.s}^{-1}$$

$$\mu = \frac{7.305.10^{-26} \times 375}{2.83 \times \pi \times \left(0.32.10^{-9}\right)^2}$$

$$\mu = 3.10^{-5}\,\text{Pa.s}$$

$$\bar{\lambda} = \frac{3\times 3.10^{-5}}{10^5}\left(\frac{\pi\times 1.38048.10^{-23}\times 293}{8\times 7.305.10^{-26}}\right)^{1/2} = 9.10^{-10}\times 2.1744.10^4$$

$$\bar{\lambda} = 1.96.10^{-5}\,\text{m}$$

We can see that $d_{pore} \ll \bar{\lambda}$.

The Knudsen diffusivity therefore comes into play:

$$D_K = \frac{375\times 10^{-9}}{3} = 1.25.10^{-8}\,\text{m}^2.\text{s}^{-1}$$

3.5.3. *Overall pore diffusivity in a porous medium*

According to Rothfeld [ROT 63]:

$$\frac{1}{D_{pore}} = \frac{1}{D_m} + \frac{1}{D_K}$$

D_m: auto-diffusivity (see Chapter 4 in [DUR 16a])

If we express the diffusivity no longer in relation to the area of the pores but to the overall area of the porous substance, then we need to take account of the porosity ε of that substance, and of the tortuosity t of the pores, which is the ratio of their true length to the thickness of the porous body. The practical diffusivity is then:

$$D_{practical} = \frac{\varepsilon}{t^2} D_{pore}$$

The factor t^2 varies from 2 to 3 and, more rarely, up to 4.

3.5.4. *Surface diffusivity [MIY 01]*

We shall suppose we have a parallel contribution, i.e. an additive contribution for diffusions:

– in the fluid filling the pores D_{pore};

– on the wall of the pores D_s

$$D = D_{pore} + KD_s$$

K: equilibrium constant: dq/dc.

In their two publications, the authors propose different ways of calculating D_s. For example:

$$D_s = D_m \exp\left(-\frac{E_r}{RT}\right)$$

where:

$E_r = 0.30 Q_{ist} + Q_0$ (their equation 15);

Q_{ist}: isosteric heat: $J.kmol^{-1}$;

E_r : activation energy: $J.kmol^{-1}$.

3.5.5. *Diffusion in pellets (or tablets or indeed spheroids)*

These pellets are composed of accumulated masses of microcrystals. The pore diameter in the microcrystals is of the order of a nanometer, whether in the case of molecular sieves or activated charcoal sieves. In the latter case, the pores are fissures between plates of graphite.

The diameter of the pores, which are the passages between microcrystals, is around 50 nm. These pores are macropores.

In two publications (both of which readers would do well to buy), Jury [JUR 77, JUR 78] gives the expression of the diffusivity of the agglomerate as a function of that in the dispersed phase and of that in the continuous phase surrounding the microcrystals (his equation 47). This expression is of the form:

$$D_{agg} = \frac{1}{\dfrac{a_0}{D_c} + \dfrac{a_1}{D_d}} + a_2 D_c$$

D_c and D_d are the diffusivities in the continuous and dispersed phases.

3.5.6. *Differential equation internal to solid particles*

Over a section with abscissa z, all particles are in an identical state. We shall suppose them to be spherical, with radius r.

In a shell whose volume is $4\pi r^2 dr$, the variation in quantity of solute present is, according to Fick's law:

$$4\pi r^2 \Delta r \frac{\partial q}{\partial t} = 4\pi D_s \left[-r^2 \frac{\partial q}{\partial r} + r^2 \frac{\partial q}{\partial r} + \Delta r \frac{\partial}{\partial r}\left[r^2 \frac{\partial q}{\partial r} \right] \right]$$

$$\frac{\partial q}{\partial t} = D_s \frac{1}{r^2} \frac{\partial}{\partial r}\left(r^2 \frac{\partial q}{\partial r} \right)$$

The notations are:

D_s : diffusivity of the solute in the solid particle: $m^2.s^{-1}$;

q : concentration of solute in the solid phase: $kmol.kg^{-1}$ or $kg.kg^{-1}$.

To render the equation dimensionless, or rather, homogeneous with respect to q, we set:

$$\tau = \frac{t}{Lu} \qquad \rho = \frac{r}{R} \qquad Pé_s = \frac{uR}{D_s}$$

Therefore, we have the equation:

$$\frac{\partial q}{\partial \tau} = D_s \frac{1}{\rho^2} \frac{\partial}{\partial \rho}\left(\rho^2 \frac{\partial q}{\partial r} \right) \qquad [3.13]$$

NOTE.–

The right-hand side of this equation [3.13] is, apart from the coefficient D_s, equal to the Laplacian of q in spherical coordinates. For particles in the shape of platelets or cylinders, it is possible to generalize the expression of the Laplacian $\Delta\rho$

$$\Delta\rho = \frac{1}{\rho^{a-1}} \frac{\partial}{\partial \rho}\left(r^{a-1} \frac{\partial q}{\partial r} \right)$$

where a is equal to 1, 2 or 3 for platelets, cylinders or spheres.

3.5.7. *Fluid–solid exchange equation*

Let us introduce the following values:

q: amount of solute in the solid phase: $kg.kg^{-1}$ or $kmol.kg^{-1}$;

c_S: concentration of solute in the solid phase: $kg.m^{-3}$ or $kmol.m^{-3}$.

We of course have:

$$\rho_1 q = c_s$$

At the surface of an isolated particle, we can write (Fick's law):

$$-D_s \frac{\partial c_s}{\partial r} = k(c^* - c)$$

When we multiply the two numbers by R, we find:

$$\left.\frac{\partial c_s}{\partial \rho}\right|_{\rho=1} = Bi\,(c^* - c) \qquad [3.14]$$

where:

$$\rho = \frac{r}{R} \text{ and } Bi = \frac{kR}{D_s}$$

Equation [3.14] expresses the *coupling* between the fluid and the solid particles.

3.5.8. *Simplified fluid–particle exchange*

To use the two-film method, we must assume that the profile of concentration inside the particle and near to its surface is shown by Figure 3.2.

In the figure, the curve A shows the true profile, and the straight lines B represent the fictitious profile.

The flux of material is, per unit surface:

$$k_f\left(c_\infty - c_I\right) = k_s\left(q_I - \bar{q}\right)$$

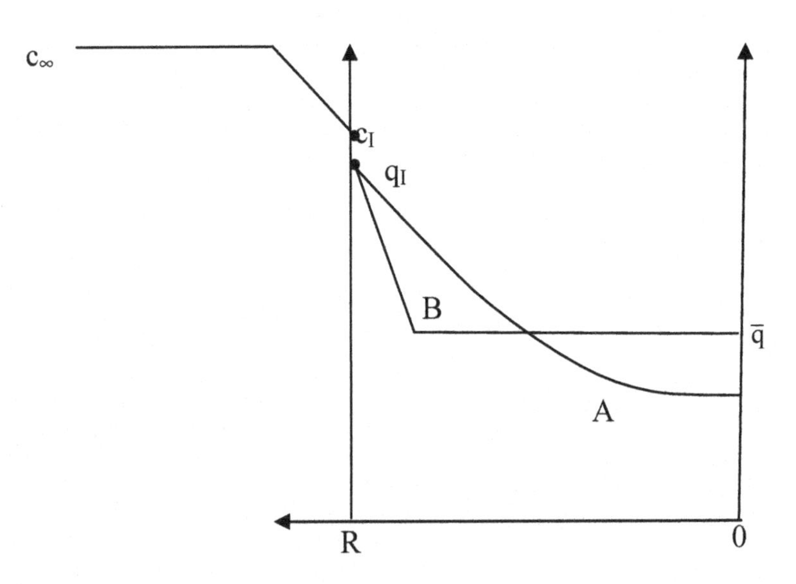

Figure 3.2. *Concentration profile*

If we take the ratio $q_I/c_I = m$, we can write:

$$m\frac{\left(c_\infty - c_I\right)}{\frac{m}{k_f}} = \frac{q_I - \bar{q}}{\frac{1}{k_s}} = \frac{mc_\infty - \bar{q}}{\frac{m}{k_f} + \frac{1}{k_s}} = \frac{mc_\infty - \bar{q}}{\frac{1}{K}} = K\left(mc_\infty - \bar{q}\right)$$

Let us stress the fact that we must not write $q^* = f(c_\infty)$, using the isotherm, but instead we must write:

$$q^* = c_\infty \frac{q_I}{c_I} = mc_\infty \quad \text{where} \quad q_I = f(c_I)$$

However, the calculation is easy if the isotherm is nearly a straight line. Otherwise, we must work in iterations. Similarly, if the resistance on the side of the fluid is negligible, then we can write the flux density:

$$k_s a_s = \frac{15D_s}{R^2} \qquad \text{(see [GLU 55])}$$

a_S: specific area of a particle (in relation to its volume): $m^2.m^{-3}$

$$a_s = \frac{3}{R}$$

k_s and k_f : transfer coefficients: $m.s^{-1}$.

The flux density is then:

$$k_s a_s \left(q^* - \overline{q}\right)$$

3.6. Adsorption between a mass of liquid and a mass of solid

3.6.1. *A simple situation*

Bring an absorbent into contact with a liquid of large volume, supposing that the *isotherm is linear. The concentration of the liquid will remain constant.* The agitation will be intense so that the *resistance of the liquid film is negligible.*

We shall set:

$\overline{q}$: mean concentration in the particle at time t: $kg.m^{-3}$;

q_0: uniform concentration in the particle at time zero: $kg.m^{-3}$;

q^*: concentration in the particle at equilibrium with the liquid: $kg.m^{-3}$.

In addition, we shall suppose that the volume of liquid is very great, so that, during the course of its adsorption to the solid, the concentration of adsorbate in the liquid does not vary.

During the operation, the surface of a particle (supposed to be spherical, with radius R) remains at equilibrium with the liquid.

The variation of $\overline{q}$ as a function of time is then given by the well-known relation:

$$\frac{m}{m_\infty} = \frac{q - q^*}{q_0 - q^*} = 1 - \frac{6}{\pi^2}\sum_{n=1}^{\infty}\frac{1}{n^2}\exp\left(-\frac{n^2\pi^2 D_s t}{R^2}\right) \qquad [3.15]$$

If m/m_∞ is greater than 0.7, we can retain only the first term and write:

$$1 - \frac{m}{m_\infty} = \frac{6}{\pi^2} \exp\left(-\frac{\pi^2 D_s t}{R^2}\right)$$

If m/m_∞ is less than 0.3, we can write:

$$\frac{m}{m_\infty} = \frac{6}{\sqrt{\pi}} \left(\frac{D_s t}{R^2}\right)^{1/2}$$

This relation was verified by Dedrick and Beckmann [DED 67] for the adsorption of impurities in water to activated charcoal.

NOTE.–

Given that equation [3.15] is slow to converge on a solution, it is possible to replace it with (see [RUT 84]):

$$\frac{nR}{\sqrt{D_s t}} \frac{m}{m_\infty} = 6\left(\frac{D_s t}{R^2}\right)^{1/2} \left[\frac{1}{\sqrt{\pi}} + 2\sum_{n=1}^{\infty} \text{erf}\left(\frac{nR}{\sqrt{D_s t}}\right)\right] - 3\frac{D_s t}{R^2}$$

3.6.2. *The volume of liquid is limited [HUA 73]*

We shall suppose that the *isotherm is linear*, that the liquid volume is limited and that the agitation is sufficient that *the resistance of the liquid film can be treated as negligible*. As the volume of the liquid is limited, *its concentration would be variable.*

The material balance is written:

$$V_L\left(c_0 - c_\infty\right) = V_S\left(q_\infty - 0\right) \;\left(\text{with } q_{t=0} = 0\right) \qquad [3.16]$$

The concentrations c_∞ and q_∞ in the liquid and in the solid are reached after an infinite period of time, and correspond to equilibrium.

$$q_\infty = Kc_\infty \qquad [3.17]$$

From equations [3.16] and [3.17], we derive:

$$q_\infty = \frac{V_L c_0}{V_S + \frac{V_L}{K}} \text{ and } c_\infty = \frac{V_L c_0}{V_S K + V_L}$$

Let us set:

$$\alpha = \frac{V_L}{K V_S} \text{ and } F(t) = \frac{q(t)}{q_\infty}$$

The solution cited by Huang and Li [HUA 73] is written:

$$F(t) = 1 - \sum_{n=1}^{\infty} \frac{6\alpha(1+\alpha)}{9 + 9\alpha + g_n^2 \alpha^2} \exp\left(-g_n^2 \frac{D_s t}{R^2}\right)$$

g_n is defined by:

$$\frac{\text{tg } g_n}{g_n} = \frac{3}{3 + \alpha g_n^2}$$

(This solution had been given by Crank, [CRA 56]).

R: radius of spherical particles: m;

Ds: diffusivity in the particles: m^2.s-1;

t: time: s;

K: equilibrium coefficient: dimensionless.

3.6.3. *Integration of the resistance of the liquid film*

Huang and Li [HUA 73] give the solution. They determine the parameter g_n by using the equation:

$$\frac{\text{tg } g_n}{g_n} = \frac{3\xi - \alpha g_n^2}{(\xi - 1)\alpha g_n^2 - 3\xi}$$

F(t) is then given by:

$$F(t)=1-\sum_{n=1}^{\infty}\frac{6\xi^2(1+\alpha)}{\left(\frac{9}{\alpha}+\alpha g_n^2+9\right)\xi^2-(6+\alpha)g_n^2\xi+\alpha g_n^4}\exp\left(-g_n^2\frac{D_s t}{R^2}\right)$$

$$\xi=\frac{R\beta}{D_s K}$$

β: coefficient of fluid–particle transfer: $m.s^{-1}$.

The coefficient β plays a role in the transfer relation between the liquid and the solid particles:

$$D_s\frac{\partial q}{\partial r}\Big|_{r=R}=\beta\left(c-\frac{q}{K}\right)$$

This relation has been perfectly borne out by experience. Indeed, the isotherm has been shown to be linear both in the case of calculation and for products undergoing testing.

3.6.4. *The isotherm is not linear*

The flux density of diffusion is:

$$N=-Dq\frac{d\text{Ln}\,\gamma q}{dz}$$

The derivative dLn γq/dz is the mobility of the molecules.

Obviously, we need to know the function γ(q), which we determined for a Langmuir isotherm.

The only way of working is to integrate the diffusion equation numerically for a given function γ(q).

3.6.5. *Bi-dispersed porosity [RUC 71]*

Here we suppose that the solid phase is divided into macrostructures of spherical form, composed of microspheres.

The macrostructures are characterized by a radius r_a and a diffusivity D_a. Microspheres are also characterized by their radius r_i and a diffusivity D_i. We also take a value for the porosities ε_a and ε_i.

The authors give the variation of the solute absorbed as a function of time. The isotherm is supposed to be linear, so the diffusion equations are simple.

4

Liquid–Solid Extraction and Washing of a Divided Solid by a Liquid

4.1. Fundaments of extraction and liquid–solid washing

4.1.1. *Terminology*

Apart from the operations of crystallization, chemical precipitation and pure-and-simple dissolution, there are two types of operations of material transfer between a liquid and a divided solid:

– extraction: in this case, the noble species passes from the solid in the solvent. At the output from the installation, the solution of the noble species in the solvent is called the extract and the depleted solid is called the residue. This operation is primarily used in food industries. Let us specify that the residue may be press cake, used coffee grounds, beet pulp or bagasse, which is the residue from sugar cane;

– washing: we also refer to this as leaching, elutriation or indeed lixiviation. The term *elutriation* comes from the Latin *elutrio*: “I rinse”. The term *lixiviation* also stems from the Latin, *lixivium*: “lye”. In the operation of washing, the noble product is the insoluble solid, which has been stripped of an impurity by the action of the solvent. Unlike extraction, the operation of washing is primarily used in the treatment of ores and crystallization.

4.1.2. *Measuring the total soluble and the extractible soluble products*

These two measurements need to be carried out on an industrial sample taken at the inlet to the extraction, so the measurements are performed on a material that has undergone mechanical and possibly hydrothermal treatments.

The extractible soluble can be measured as follows:

1) Over a time period equal to 5 times the residence time of the product in the extraction chamber, with the sample being brought into contact with the solvent by gentle agitation (simply to avoid setting). The solvent shall be the same as that which is used in the industrial installation.

2) After pressing of the product, it is dispersed anew in the pure industrial solvent with gentle agitation for the same period of time as before. The product is then pressed again.

3) We repeat the operation once again, but this time for only an hour, again using the industrial solvent, and we press it.

4) To obtain the mass M_{sU} of soluble corresponding to 1 kg of product coming into the extraction process, the three solvents are mixed and subjected to vaporization in order to recover and weigh the soluble.

5) Having isolated the dry soluble, we determine its true density ρ_{sU} using a pycnometer and a non-dissolvent liquid.

6) For 1 kg of product entering into the extraction process, the volume of soluble is then:

$$V_{sU} = M_{sU}/\rho_{sU}$$

The total soluble is measured in the same way but, beforehand, the sample will have undergone ultra-fine milling (product smaller than 5μm). The milling can thus be carried out:

1) dispersion in the solvent,

2) treatment in an attrition mill,

3) emptying of the mill,

4) rinsing of the mill and balls with pure solvent,

5) mixing of the rinse solvent and dispersion,

6) return to step 1.

By measuring the extractible soluble and the total soluble, we are able to gage the effectiveness of the extraction device, and also to reach the inevitable loss because, in industrial practice, it is out of the question to mill sugar cane, coffee grains or oleaginous grains smaller than 5 µm.

Hereafter, as we shall focus on the performances of the extractors, the concentration of soluble in the product entering the extractor will be the concentration of extractible soluble.

4.1.3. *Accessible porosity*

The accessible porosity is the fraction of volume of the particles which may be occupied by the solvent–soluble mixture.

1) We immerse a given mass of particles in the solvent at the extraction temperature for 30 minutes with gentle agitation so that any swelling of the particles has the time to occur.

2) We separate the solvent and the particles by moderate pressing so as not to empty the pores of the particles. From this, we deduce the mass M_p of particles impregnated per kg of product coming into the extraction device.

3) Using a pycnometer and a non-wetting liquor, we measure the density ρ_p of the impregnated particles. From this we deduce the volume $V_p = M_p/\rho_p$ of particles impregnated per kg of product entering the process.

4) Given that biological tissues include a large proportion of aqueous solutions, it is helpful to distinguish three cases:

– the solvent is not miscible with water and does not cause swelling of the solid material. During the course of the extraction, it simply takes the place of the soluble and the accessible porosity would be:

$$\varepsilon_A = \frac{V_{SU}}{V_p}$$

– the solvent is still not miscible with water, but can cause swelling of the solid material. The fraction of solvent used for swelling does not play a part in the extraction, and we again have:

$$\varepsilon_A = \frac{V_{SU}}{V_p}$$

– the extraction is performed with water:

$$\varepsilon_A = \frac{V_{SU}+V_H}{V_p} \quad \text{with} \quad V_H = M_H/1000$$

M_H is the mass of water present in the mass M_p of impregnated particles. It is measured by oven drying. This case pertains to sugar cane, sugar beet and instant coffee, whilst sections 1 and 2 pertain to oleaginous compounds.

4.1.4. *Equilibrium coefficient*

At equilibrium, the composition of the liquor outside of the particles must be equal to that of the liquor filling its pores.

The concentration of soluble expressed in relation to the volume of the particles is:

$$q = \varepsilon_A c$$

where c is the concentration of the external liquor.

By definition, the equilibrium coefficient is:

$$m = \frac{c}{q} = \frac{1}{\varepsilon_A}$$

By immersing samples of the product in varying quantities of solvent, we vary the concentration of the external liquor, which enables us to verify whether m is indeed independent of this concentration.

If we see that m varies, then we set:

$$m = \frac{\beta(c)}{\varepsilon_A}$$

If $\beta > 1$, the particles are less rich in solvent than expected, meaning that the solid matrix has a preferential affinity for the soluble.

If $\beta < 1$, the particles are richer in solvent than expected, which means that the solid matrix has a preferential affinity for the solvent.

The proposed method to measure the extractible soluble is such that $1/\varepsilon_A$ precisely represents the slope at the origin of the equilibrium curve. It is only for concentrations significantly different to zero that $\beta(c)$ may be different to 1.

4.1.5. *Shape and dimension of particles*

1) We immerse a given mass of particles in the solvent, as we did to determine the accessible porosity.

2) We then wring the particles out.

3) We examine them under the microscope to determine their shape (spheroids, fibers, plates).

4) We percolate the solvent through the particles. Then we measure the volume of the "cake" and, knowing the mass of the particles, we deduce their true volume by division by ρ_p. Thus, we are able to obtain the porosity of the cake.

The permeability is expressed by:

$$K = \frac{\mu V Z}{\Delta P} = \frac{\varepsilon^3}{4.17\sigma^2 (1-\varepsilon)^2} \qquad 4.17=150/36 \qquad [4.1]$$

Knowing the viscosity μ of the solvent, its velocity in an empty bed V, the thickness Z of the cake and the pressure drop ΔP on crossing the cake, we deduce σ, which is the external surface of the particles in relation to their volume.

5) By knowing σ, we are able to obtain the characteristic dimension of the particles:

$$\sigma = \frac{6}{d_s} = \frac{4}{d_c} = \frac{2}{e_p} \qquad [4.2]$$

Respectively, d_s and d_c are the diameter of the sphere and the equivalent cylinder and e_p is the thickness of the equivalent plate.

From the point of view of diffusion, once we have determined d_c or e_p, we need to calculate the radius R_s of the equivalent sphere for the extraction, using the solutions to the diffusion equation. In the case of liquid–solid extraction, we can consider that the Biot number is high because the resistance of the film surrounding the particles is negligible in comparison to that due to the slowness of diffusion within the particles. This means that, at the surface of the particles, the liquid has the same composition as the liquor in which the particles are bathed.

If we take only the first term of the serial expansion of the solutions to the diffusion equation for the sphere, the cylinder and the plate, we obtain:

$$\text{sphere}: \ \bar{c} = A \exp\left[-\pi^2 \frac{d\tau}{R_S^2}\right]$$

$$\text{cylinder}: \ \bar{c} = B \exp\left[-(2.405)^2 \frac{d\tau}{R_C^2}\right]$$

$$\text{plate}: \ \bar{c} = C \exp\left[-\left[\frac{\pi}{2}\right]^2 \frac{d\tau}{(e_p/2)^2}\right]$$

For a time of around half an hour, we should have:

$$\frac{\pi^2}{R_s^2} = \frac{(2.405)^2}{R_c^2} = \frac{(\pi/2)^2}{(e_p/2)^2} = \frac{\pi^2}{e_p^2}$$

For the cylinder:

$$R_s = R_c \times \frac{\pi}{2.405} = 1.3\, R_c \qquad [4.2b]$$

For the plate:

$$R_S = e_p$$

4.1.6. *Global transfer coefficient in chromatography*

This coefficient must express the influence of the resistance of the fluid film surrounding the particle and that of the diffusion inside of the particle.

The transfer at the surface of a spherical particle with radius R_s is expressed by:

$$D\frac{\partial q}{\partial R}(R_s, x, \tau) = \beta\left[c^*(x,\tau) - c(x,\tau)\right]$$

D: diffusivity of the soluble in the particle: $m^2.s^{-1}$;

β: transfer coefficient: $m.s^{-1}$;

x: abscissa along the displacement of the fluid: m;

τ: time: s.

In addition:

$$D\frac{\partial q}{\partial R}(R_s, x, \tau) = \beta\left[c^*(x,\tau) - c(x,\tau)\right]$$

c^*: concentration of the solution outside of the particles: $kg.m^{-3}$ or $kmol.m^{-3}$;

q: concentration of particles: kg or kmol per m^3 of particles;

m: equilibrium coefficient.

We can bring in the mean concentration in the particles:

$$\bar{q} = \frac{3}{R_s^3}\int_0^{R_s} qR^2 dR$$

If we adopt this hypothesis, the transfer at the surface of the particle is written as follows, as we shall see later on:

$$\frac{\partial \bar{q}}{\partial \tau} = -K_c a\left(c^* - c\right)$$

where:

$$c^* = m\bar{q}$$

Within the grains, diffusion is expressed by:

$$\frac{\partial \bar{q}}{\partial \tau} = \frac{1}{R^2}\frac{\partial q}{\partial R}\left[DR^2\frac{\partial q}{\partial R}\right]$$

4.1.7. *Equivalence between chromatography and saturation of a fixed bed*

By studying the dynamics of saturation of a fixed bed that is initially free of soluble, we can accept:

– either that the concentration of the incoming liquid is a step function of time;

– or that the concentration of the incoming liquid is a simple pulse.

The response to a step function is typically called the penetration curve (or rupture curve) and the response to a pulse is a peak which is the chromatographic response.

However, a Dirac function (perfect pulse) is the derivative of a Heaviside function (perfect step function) for a linear system so that the penetration curve is the integral of the chromatographic response.

Liquid–solid extraction in a fixed bed can be considered the opposite of adsorption, because we are stripping the substance out of a bed that is initially saturated. However, for a constant equilibrium coefficient m, the concentration of effluent in the case of depletion is the complement of the penetration curve on enrichment. Indeed, in adsorption, the concentration of the effluent rises from zero to c_F and, in extraction, it falls from mq_o to zero.

The above means we can legitimately write:

K_c: global transfer coefficient: $m.s^{-1}$

In the discussion below, we shall no longer include the bar above q.

By way of considerations implementing the Laplace transformation, Haynes and Sarma [HAY 73] showed, by identifying the 2nd-order moment of the chromatographic response, that (their equation 23):

$$\frac{1}{K_c a} = \frac{R_s}{3\beta} + \frac{R_s^2}{15D}$$

β: transfer coefficient at the surface of the particle: $m.s^{-1}$;

R_S: radius of the equivalent sphere of the particle: m;

a: surface area of transfer expressed in relation to the volume of the particle: m^{-1}.

In liquid–solid extraction, β is large enough for the first term to be negligible in comparison to the second, and we simply write:

$$\frac{1}{K_c a} = \frac{R_s^2}{15D} \qquad [4.3]$$

R_S is the radius of the equivalent sphere of the particles for diffusion.

a is the interfacial area of transfer expressed per m^3 of solid, whilst a_L is expressed per m^3 of the bed.

$$a_L = (1 - \varepsilon)a$$

We shall use a_L for percolation extractors.

D is the diffusivity of the soluble in the particles and can be obtained by:

$$D = D^* \frac{\varepsilon_A}{t^2}$$

ε_A is the accessible porosity of the particles and t the tortuosity for percolation within the particles. As an initial approximation, it is sufficient to take:

$$t = 1.45$$

D^* is the diffusivity of the soluble in the solvent and can be estimated by the Stokes–Einstein equation (which is valid for molecules that are not too small):

$$D^* = \frac{k_B T}{6\pi\mu r} \qquad [4.4]$$

k_B: Boltzmann's constant: 1.38×10^{-23} J.K^{-1};

μ : viscosity of the liquor: Pa.s;

r : molecular radius of the solute: m.

It is possible to quickly estimate r using the relation:

$$\frac{4\pi r^3}{3} = \frac{M}{\rho_{sU} N_A}$$

M: molar mass of the solute: $kg.kmol^{-1}$;

N_A: Avogadro's number: 6.02×10^{26} $molecule.kmol^{-1}$;

ρ_{sU}: density of the soluble in the solid state: $kg.m^{-3}$.

4.1.8. *Number of transfer units on the side of the extract*

We refer here to a classic calculation, in the case where the product and the extract circulate in a perfect countercurrent flow.

A balance around the device is written:

$$Q_E(c_o - c_i) = Q_R(q_i - q_o)$$

c and q: concentrations of soluble in the extract and in the product, measured in $kg.m^{-3}$ or in $kmol.m^{-3oi}$;

Q_E and Q_R: flowrates of extract and residue: $m^3.s^{-1}$.

The indices e and s refer respectively to the inlet and the output.

The ideal number of transfer units on the side of the extract is defined by:

$$N_{EI} = \int_{ci}^{co} \frac{dc}{c^* - c}$$

If the equilibrium coefficient is m, the equilibrium equation is written:

$$c^* = mq$$

Finally, the ideal extraction factor is written, conventionally:

$$\varepsilon_I = m\frac{Q_E}{Q_R}$$

There are two ways of performing the calculation, depending on whether we find a balance around the end poor in soluble or around the rich end:

1) Balance around the end rich in soluble:

$$Q_E(c_o - c) = Q_R(q_i - q)$$

Hence:

$$q = \frac{Q_E}{Q_R}(c - c_s) + q_i$$

By replacing c^* with mq in the integral, we obtain:

$$N_{EI} = \int_{c_e}^{c_s} \frac{dc}{\mathcal{E}_I(c - c_s) + mq_e - c}$$

and, after integration:

$$(\mathcal{E}_I - 1)N_{EI} = \mathrm{Ln}\left[\frac{mq_i - c_s}{c_i(\mathcal{E}_I - 1) + mq_i - \mathcal{E}_I c_o}\right]$$

2) Balance around the end poor in soluble:

$$Q_E(c - c_i) = Q_R(q - q_o)$$

Hence:

$$q = \frac{Q_E}{Q_R}(c - c_i) + q_o$$

Therefore:

$$N_{EI} = \int_{c_u}^{c_o} \frac{dc}{\mathcal{E}_I(c - c_i) + mq_o - c}$$

and finally:

$$(\mathcal{E}_I - 1)N_{EI} = \mathrm{Ln}\left[\frac{c_i(\mathcal{E}_I - 1) + mq_o - \mathcal{E}_I c_i}{mq_o - c_i}\right]$$

4.1.9. *Extractors and washers*

In 1965, Rickles put forward a detailed list of liquid–solid extraction procedures used, whether the solid is organic or inorganic. From a more fundamental point of view, he laid out the theory of *penetration*, that of the *limiting film* and that based on *capillary flow*. Hereafter, we shall go into detail about the most commonly-used processes, and distinguish load-based devices and continuous devices.

1) Devices operating by successive loads

The product is amassed in a vat whose wall is heated by vapor. The solvent (often water) then fills the device to capacity. At the bottom is a sheet of perforated metal, divided into sectors which can be removed to clean out the debris that has fallen into the bulbous bottom.

The product may remain in this sort of infuser between 8 hours and 5 days, depending on the case. This extractor is used to extract the active substance from the leaves or flowers. There need be no agitation, apart from the convection currents in the solvent.

For draining, we subject the infuser to pressure from compressed air (0.5 to 1 bar) and the product exits through the manhole at the bottom.

A very different system is used to wash ores (such as gold- or uranium ore). The system is a vertical tank, moderately agitated by the injection of air into an axial drawing tube. The dispersion of the ore in the wash liquor rises into the tube because it is entrained by the air, and falls back down at the periphery. The inside of the device at the bottom is conical and the top can be left open to the air.

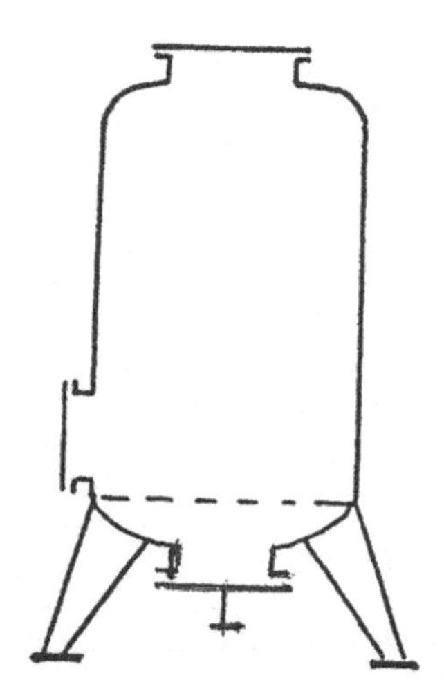

Figure 4.1. *Discontinuous infuser*

Such devices, which are large, are used for slow operations such as the crystallization of alumina (the operation takes 8 days). These devices can run continuously.

2) Immersion continuous devices

For the treatment of ores, it is possible to design a series of stages each composed of a stirred vat (the mixer) and a gravity settler.

Quite unlike the washing of ores, the washing of the crystals produced by the chemical industry is a very quick operation, because they are simply coated in the parent liquor on the surface. It is important to cleanse them of it, because that liquor may contain impurities. Therefore, for crystals whose size is greater than or equal to 200 μm, we use a column where the crystals are fluidized by the wash liquor.

A drawback to the fluidized column is that it requires a high flowrate of liquor, and if we reduce the size of the product so as to decrease the liquor flowrate, the device tends to become a gravity decanter whose rake, i.e. the height/diameter ratio, is too slight. The device then functions as a single mixer/decanter stage, no longer with a counter-current.

The screw extractor (Figure 4.2) or inclined-blade extractor then offers a solution, because the screw or blades take care of transporting the product through the extended part of the device.

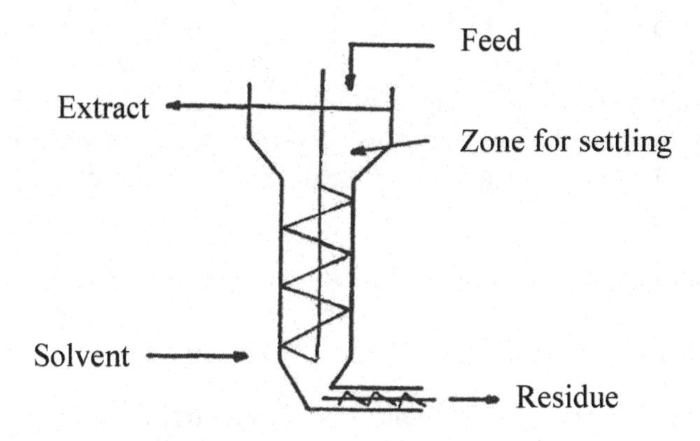

Figure 4.2. *Screw extractor*

The diffusion batteries used for instant coffee are of extremely simple design, because they are simply a series of vats where the product is placed in a fixed bed. With a periodicity equal to the cycle time, the order of the vats is modified relative to the circulation of the extract.

In the past, these batteries were used for sugar beet. For this purpose, the beet was cut up into "cosettes", having the form of tiles with a V-shaped cross-section whose thickness ranged from 1 to 2.8 mm. This low thickness favored the extraction of sugar by diffusion. In addition, the V-shaped section ensures significant porosity of the bed and facilitates the flow of the juice. Today, sugar beet is still cut into cosettes, but extraction from those cosettes is a continuous process in devices designed by sugar companies.

3) Watering continuous devices (so-called percolation devices)

The product moves slowly along a transporting belt, above which are extract sprayers. For each spray system there is a corresponding "section" and, in relation to the advancement of the solid, the extract which has passed through it (and also through the perforated transporting belt) is collected in a run-down funnel, recovered by a pump and sprayed on the next section upstream. This configuration is similar to the countercurrent setup.

4) Below are some examples of simulation for devices irrigating oleaginous grains and sugar cane, and the example of a diffusion battery for instant coffee.

These examples use the method put forward by Spaninks and Bruin [SPA 79]. They are not intended to reflect any particular installation, but are based on numerical values which can be viewed as reasonable. In particular, to simplify the calculations, the density of the species in general is taken as equal to 1000 kg.m^{-3}, because we are dealing with food products.

4.2. Hydrodynamics of percolation continuous extractors

4.2.1. *Principle*

The product moves slowly along a perforated transporting belt. Ten or so sprayers are arranged in a series along the belt and moisten the top of the bed of product.

We shall see that with this device, it is possible to use relatively little solvent, which is advantageous for later distillation or evaporation: there is less solvent to be vaporized, and hence a saving in terms of energy. However, there is a lower bound to the flowrate of solvent if we want the transfer potential $(c^* - c)$ to retain an economical value.

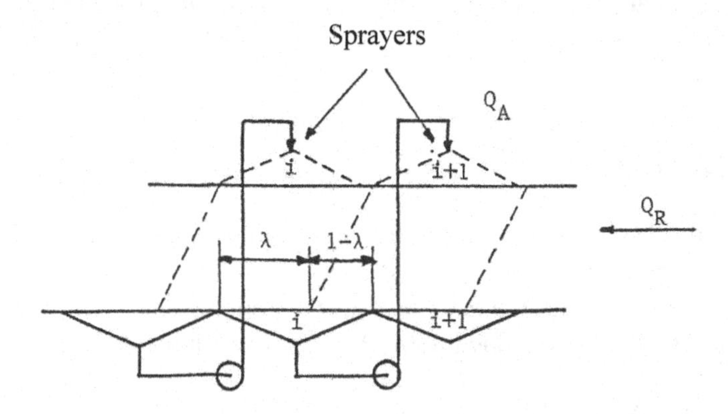

Figure 4.3. *Percolation device*

Because the solid moves horizontally, the trajectory of the liquid is inclined on the vertical, but the rundown tanks are staggered in relation to the solid, so that tank i recycles back into section i + 1 the fraction $1 - \lambda$ of the liquid coming from it.

Let Q_A represent the irrigation flowrate of a sprayer. The fraction λQ_A of that flowrate corresponds to the flowrate of extract Q_E, which circulates in a countercurrent flow of the product and the fraction $(1 - \lambda)\, Q_A$ cycles around the section. We can define a recirculation rate R:

$$R = \frac{1}{1-\lambda}$$

Recirculation has a positive impact on the interfacial area of material transfer. Indeed, by thus increasing the irrigation flowrate (i.e. the percolation), we obtain the following results:

– if the product has low porosity, we approach operation in the soaked regime, where the product is completely immersed in the extract. The interfacial area is maximal;

– if the product is highly porous, the solid will simply be irrigated, but more so when the flowrate Q_A is higher.

We shall establish the expression of the velocity in an empty bed of the liquid for flooding. Flooding corresponds to the flowrate of liquid for which the level of that liquid in the product reaches the surface of the bed. Using two examples – an oleaginous grain and sugar cane – we shall see what rate of recirculation we can expect.

4.2.2. *Flooding rate*

The only driving force which makes the liquid traverse the product is a volumetric force due to the field of gravity. If A is the horizontal surface area of a section and Δx the thickness of a horizontal slice, the force which is exerted on the liquid contained in the elementary volume $A\Delta x$ is:

$$\Delta F = A\rho_L \varepsilon \Delta x g$$

ρ : density of the liquid: $kg.m^{-3}$;

ε : porosity of the product;

g : acceleration due to gravity: 9.81 $m.s^{-2}$.

This force has two effects:

– neutralizing the viscous friction;

– accumulating enough hydrostatic pressure ΔP_H to overcome the resistance of the perforated belt supporting the product.

$$\frac{\Delta F}{A\Delta x} = \rho_L \varepsilon g = \frac{\mu V}{K} + \frac{\Delta P_H}{\Delta x}$$

μ : viscosity of the liquid: Pa.s;

K : permeability of the bed: m^2;

V : velocity in an empty bed of the liquid: $m.s^{-1}$

$$V = Q_A/A$$

That is to say:

$$\frac{\partial P_H}{\partial x} = \rho_L \varepsilon g - \frac{\mu V}{K}$$

With the boundary conditions:

$$x = 0 \quad P_H = 0$$

$$x = Z \quad P_H = R\mu V$$

x is the score counted downwards from the surface of the bed, R is the resistance of the support grid and Z is the thickness of the bed of product.

By integrating with x ranging from zero to Z, we find the expression of the flooding speed V_E:

$$V_E = \frac{\rho_L \varepsilon g Z}{\mu\left(R + \frac{Z}{K}\right)} \qquad [4.5]$$

From this expression, we can see that if we introduce liquid at a rate less than V_E, only the lower fraction of the bed will be soaked by the liquid.

On the other hand, if the irrigation flowrate is greater than V_E, a thickness H_L of liquid will form above the product. Thus, at the top of the product, the hydrostatic pressure will no longer be 0, but instead will be equal to $\rho_L g H_L$.

In light of this new boundary condition, we obtain, by integration:

$$V = \frac{\rho_L \varepsilon g (Z + H_L)}{\mu\left(R + \frac{Z}{K}\right)}$$

However, percolation extractors never work in this way.

NOTE.–

Strictly speaking, if we wanted to maintain an flooding rate V/V_E constant for all the sections, we would have to decrease the recirculation as the concentration of the extract increases, i.e. as the viscosity increases. We would also have to take account of any variations in temperature along the length of the device, because temperature has a major influence on viscosity.

NOTE.–

A definite advantage to recirculation is that it mitigates the harmful effect of preferential flows in the bed (channeling).

EXAMPLE 4.1 (Oleaginous grains).–

After milling and rolling on smooth cylinders, the flour derived from oleaginous grains is presented in the form of flakes whose thickness is around 0.2 mm. By definition, those chips are flat. If they were as rigid as paraffin crystals (which are plate-like in form), their porosity ε would be between 0.85 and 0.95 but, because they are relatively plastic and deformable, their porosity is lesser. We can estimate it as around 0.6.

As this porosity is not particularly great, we adopt a bed height of 0.5 m.

The volumetric flowrate of solvent is taken as equal to half the true volumetric flowrate of flour.

For the calculation of the permeability, we have (see equation [4.2]):

$$d_s = 3e_p = 0.6.10^{-3}\ \text{m}$$

Hence (equation [4.1]):

$$K = \frac{0.6^3 \left(0.6.10^{-3}\right)^2}{150(1-0.6)^2} = 3.2.10^{-9}\,\text{m}^2$$

The resistance of the belt is supposed to be slight, equal to $10^6\ \text{m}^{-1}$; the viscosity of the miscella (a mixture of solvent and oil, from the Latin *miscellus*, meaning "mixture") is taken as equal to 0.6×10^{-3} Pa.s at 70°C. The rate of flooding is then (with a density of 900kg.m^{-3} for the miscella) (see equation [4.5]):

$$V_E = \frac{900 \times 0.6 \times 9.81 \times 0.5}{0.6.10^{-3}\left[10^6 + \dfrac{0.5}{3.2.10^{-9}}\right]}$$

$$V_E = 0.028\ \text{m.s}^{-1}$$

If a section is 2 m in length, if we let ℓ represent the breadth of the bed of product and if the velocity of the belt is 0.6 m.mn^{-1}, the flowrate of flour will be:

$$Q_R = (1 - 0.6) \times 0.5 \times l \times 0.01 = (0.02 \times l)\ m^3.s^{-1}$$

The flowrate of solvent is $0.5Q_R$. The velocity in an empty bed for the percolation will be:

$$\frac{0.5Q_R}{2 \times l} = \frac{0.50 \times 0.02}{2} = 0.005\ m.s^{-1}$$

If we want to reach 70% of the rate of engorgement, we need to accept the following rate of recirculation:

$$R = \frac{0.028 \times 0.7}{0.005} = 4$$

EXAMPLE 4.2 (Sugar cane).–

After milling with a hammer mill, sugar cane, which then becomes known as bagasse, appears as short fibers a few millimeters long, which can be assimilated to cylinders whose mean diameter is 0.8 mm.

The porosity of fibrous masses is generally greater than 0.95 but, here, as the fibers are short, the porosity will be lesser, and we accept the value 0.93.

Sugar cane contains around 15% saccharose, and we choose a volumetric flowrate of juice (the solvent) equal to the flowrate of bagasse. Half the flowrate would give a solution of 30% sugar, whose high viscosity would impede the transfers of material and heat.

With a permeable product like bagasse, the height of the bed can be taken as equal to 1.5 m.

To calculate the permeability, the diameter of the spherical particle equivalent to the cylindrical fibers will be:

$$1.5 \times 0.8.10^{-3} = 1.2.10^{-3}\ m$$

Hence, the permeability is (see equation [4.1]):

$$K = \frac{0.93^3 \left(1.2.10^{-3}\right)^2}{150\left(1-0.93\right)^2}$$

$$K = 1.57.10^{-6}\ m^2$$

The viscosity of the juice will be taken as equal to 1.3×10^{-3} Pa.s, and the resistance of the transporting belt will be equal to 10^6 m^{-1}. The velocity at engorgement is then (see equation [4.5]):

$$V_E = \frac{1000 \times 0.93 \times 9.81 \times 1.5}{1.3.10^{-3}\left[10^6 + \frac{1.5}{1.57.10^{-3}}\right]} = 5.38\ m.s^{-1}$$

If a section is 2 m long, if ℓ is the width of the bed and if the velocity of the belt is 0.6 m.mn^{-1}, the flowrate of solid will be:

$$Q_R = (1 - 0.93) \times 1.5 \times l \times 0.01 = (0.001 \times l)\ m^3.s^{-1}$$

The flowrate of the juice Q_E is taken as equal to Q_R. The percolation speed is then:

$$V = \frac{0.001 \times l}{2 \times l} = 0.0005\ m.s^{-1}$$

We can see that the velocity of the juice in an empty bed is negligible in comparison to the velocity at flooding. Let us agree on a recirculation rate of 15 because, beyond this value, the electricity consumption would be unreasonably high. The velocity in an empty bed for the watering would be:

$$0.0005 \times 15 = 0.0075\ m.s^{-1}$$

The soaked height Z will then be such that:

$$0.0075 = \frac{1000 \times 0.93 \times 9.81 \times Z}{1.3.10^{-3}\left[10^6 + \frac{Z}{1.57.10^{-6}}\right]}$$

$$Z = 2.10^{-3}\ m$$

In other words, the bed 1.5 m high will operate entirely in the irrigated regime; not the soaked regime.

To operate in the soaked regime, we simply need to sharply increase the hydraulic resistance of the transporting belt and surround it with vertical walls.

4.3. Performances of continuous percolation extractors

4.3.1. *Hypothesis*

We shall surmise that, in a section, the liquid is homogeneous and has a constant composition. Indeed, if the recirculation flowrate is high, as is generally the case, the composition of the liquid will vary little throughout the thickness of the bed of product.

We also adopt the hypothesis that there is no draining of the liquid between two sections. In other words, the interstitial liquid is entirely transported by the product from one section to the next.

The method used shall be that employed by Spaninks and Bruin [SPA 79b], which we shall describe more precisely than those authors did. That description will be followed up with two examples.

4.3.2. *Number of transfer units; transfer equation*

The material transfer in the section i is written:

$$Q_R \frac{dq_i}{dx} = -K_c a_L A(c_i^* - c_i) = -K_c a_L A(mq_i - c_i)$$

where x is the distance measured in the direction of progression of the product. As we can see from Figure 4.3, the index i of the sections increases in the direction of progression of the solvent. A is the vertical area of the bed of product.

Let us integrate from one vertical face of the slice i to the other:

$$\frac{mq_i - c_i}{mq_{i+1} - c_i} = \exp\left[\frac{-K_c a_L m\Omega_p}{nQ_R}\right] = \exp\left[-\frac{\mathcal{E}N_E}{n}\right]$$

Ω_p is the apparent volume of product present in the device:

$$\Omega_p = A \qquad L\varepsilon = m\frac{Q_E}{Q_R} \qquad N_E = \frac{K_c a_L \Omega_p}{Q_E}$$

N_E is the number of transfer units and n is the number of sections.

We set:

$$mq_i = c_i^* \quad \text{et :} \quad \exp\left[\frac{\varepsilon N_E}{n}\right] - 1 = \frac{1}{\alpha} \qquad [4.6]$$

Equation [4.4] becomes:

$$c_i = q_i^* - \alpha(q_{i+1}^* - q_i^*)$$

This is the transfer equation.

BALANCE OF A SECTION.–

In the section which has the index i, we can write the balance:

$$Q_E(c_i - c_{i-1}) \quad = \quad Q_E(q_{i+1} - q_i) \quad + \quad \Delta Q_E(c_{i+1} - c_i)$$

gained by liquid — lost by the solid — brought from section i + 1 by the interstitial liquid

ΔQ_E is the interstitial liquid brought by the product:

$$\Delta Q_E = V_B A \varepsilon_h S$$

ε_h is the hydrodynamic porosity and S the saturation. The saturation is equal to 1 for a soaked bed and significantly less than 1 (e.g. 0.1 or 0.2) for an irrigated bed.

We set:

$$\varepsilon = m\frac{Q_E}{Q_R} \qquad E = m\frac{\Delta Q_E}{Q_R} \qquad c_i^* = mq_i$$

The above balance is written:

$$c_{i+1}^* - c_i^* + c_{i+1}E - c_i(E + \varepsilon) + c_{i-1}\varepsilon = 0$$

Let us calculate the concentrations at the output:

In the balance equation, let us eliminate the concentrations of the liquid using the transfer equation written for c_{i+1}, c_i and c_{i-1}. We obtain:

$$-\alpha E c^*_{i+2} + (1 + E + 2\alpha E + \alpha \mathcal{E}) c^*_{i+1}$$

$$-(1 + E + \mathcal{E} + 2\alpha\mathcal{E} + \alpha E) c^*_i + c^*_{i-1}\mathcal{E}(1 + \alpha) = 0$$

This can be written:

$$c^*_{i+2} T_1 + c^*_{i+1} T_2 + c^*_i T_3 + c^*_{i-1} T_4 = 0$$

We can see that:

$$T_1 + T_2 + T_3 + T_4 = -\alpha E + 1 + E + 2\alpha E + \alpha\mathcal{E}$$

$$-1 - E - \mathcal{E} - 2\alpha\mathcal{E} - \alpha E + \alpha\mathcal{E} + \mathcal{E} = 0$$

Thus, if we seek a solution of the form, the solution $\lambda_i = 1$ works well, and the equation is written:

$$(\lambda - 1)[T_1\lambda^2 + (T_1 + T_2)\lambda + (T_1 + T_2 + T_3)] = 0$$

Let λ_2 and λ_3 represent the solutions to the second-degree equation. Then, the general solution we are looking for is:

$$c^*_i = p_1 + p_2\lambda_2 + p_3\lambda_3$$

The transfer equation [4.5] gives us:

$$c_i = p_1 + p_2[1 - \alpha(\lambda_2 - 1)]\lambda_2 + p_3[1 - \alpha(\lambda_3 - 1)]\lambda_3$$

To simplify the formula, we set:

$$g_j = 1 - \alpha(\lambda_j - 1) \quad (j = 2 \text{ ou } 3) \qquad [4.7]$$

The coefficients p_1, p_2 and p_3 are determined using the boundary conditions:

Solvent inlet, i = 0:

$$c_F = c_o = p_1 + p_2 g_2 + p_3 g_3 \qquad [4.8]$$

Product inlet, i = n + 1:

$$c_F^* = c_{n+1}^* = p_1 + p_2\lambda_2^{n+1} + p_3\lambda_3^{n+1} \quad [4.9]$$

A third relation is obtained by writing that there is no material transfer in section n + 1 and that, consequently, the composition of the liquid exiting that section is identical to the composition of the liquid which entered it:

$$c_n = c_{n+1}$$

Put differently:

$$p_2 g_2 \lambda_2^n (\lambda_2 - 1) = -p_3 g_3 \lambda_3^n (\lambda_3 - 1) \quad [4.10]$$

By solving the system of equations [4.8], [4.9] and [4.10], we find:

$$p_1 = c_F - p_3(Bg_2 + g_3)$$

$$p_2 = Bp_3$$

$$p_3 = \frac{q_F^* - c_F}{B\left(\lambda_2^{n+1} - g_2\right) + \left(\lambda_3^{n+1} - g_3\right)}$$

where:

$$B = -\frac{g_3\lambda_3^n(\lambda_3 - 1)}{g_2\lambda_2^n(\lambda_2 - 1)}$$

Thus, the output compositions can be deduced, by making:

$i = n$	c_n	(composition of the extract)
$i = 0$	$q_o = \frac{c_o^*}{m}$	(composition of the residue)

EXAMPLE 4.3 (Oleaginous grains).–

Flours treated with solvent almost always have an oil content of less than 20% (in mass) – either because this is the natural content of the grain or else, if that content is lesser, because the grain has undergone prior pressing. Let us state that the oil content is 15% in the example discussed here.

$$m = \frac{1}{\varepsilon_A} = \frac{1}{0.15} = 6.7 \qquad \varepsilon_h = 0.6 \qquad c_F = 0$$

$$E=\frac{m\varepsilon}{1-\varepsilon}=\frac{6.7\times 0.6}{1-0.6}=10\frac{Q_E}{Q_R}=0.5\,q_F=150\text{ kg.m}^{-3}$$

$$\mathcal{E}=m\frac{Q_E}{Q_R}=6.7\times 0.5=3.35 \qquad n=10$$

The length of stay of the product is half an hour.

The solute has the molar mass 200 and true density 0.8.

$$\frac{4\pi r^3}{3}=\frac{200}{800}\times\frac{1}{6{,}02.10^{26}};\text{ hence: } r=4{,}7.10^{-10}\text{ m}$$

The mean viscosity of the miscella is 6×10^{-3} Pa.s. The Stokes–Einstein formula gives us (see equation [4.4]):

$$D^*=\frac{1.38.10^{-23}\times 300}{6\pi\times 6.10^{-3}\times 4.7.10^{-10}}=7.80.10^{-11}\text{ m}^2.\text{s}^{-1}$$

$$D=\frac{D^*\varepsilon_A}{t^2}=7.8.10^{-11}\times\frac{0.15}{2}=0.58.10^{-11}\text{ m}^2.\text{s}^{-1}$$

Thickness of the flakes: 0.2 mm.

$$\frac{1}{K_c a_L}=\frac{1}{(1-0.6)}\times\frac{\left(2.10^{-4}\right)^2}{15\times 0.58.10^{-11}}=1149\text{ s}$$

$$\frac{\Omega_p(1-\varepsilon)}{Q_R}=\frac{1}{2}\text{h}=1800\text{ s, hence: }\frac{\Omega_p}{Q_R}=\frac{1800}{1-0.6}=4500\text{ s}$$

$$\mathcal{E}N_E=\frac{K_c a_L\Omega_p}{Q_E}m\frac{Q_E}{Q_R}=K_c a_L m\frac{\Omega_p}{Q_R}=\frac{4500}{1149}\times 6.7$$

$$\mathcal{E}N_E=26.24$$

According to equation [4.6]:

$$\alpha=\left[\exp\left[\frac{26.24}{10}\right]-1\right]^{-1}=0.078$$

$$T_1=-0.078\times 10=-0.78$$

$$T_2=1+10+2\times 10\times 0.078+0.078\times 3.35$$

$$T_2 = 11 + 1.56 + 0.26 = 12.82$$

$$T_3 = -(1 + 10 + 3.35 + 2 \times 0.078 \times 3.35 + 0.078 \times 10)$$

$$T_3 = -(14.35 + 0.52 + 0.78) = -15.65$$

$$T_4 = 3.35(1 + 0.078) = 3.61$$

$$-0.78\lambda^2 + 12.04\lambda - 3.61 = 0$$

$$\lambda = \frac{-12.04 \pm \sqrt{144.96 - 11.26}}{-1.56}$$

$$\lambda = 7.72 \pm 7.41$$

$$\lambda_2 = 15.13 \qquad g_2 = 1 - 0.078 \times 14.13 = -0.1$$

$$\lambda_3 = 0.31 \qquad g_3 = 1 - 0.078(0.31 - 1) = 1.054$$

$$B = \frac{1.054 \times (0.31)^{10}(0.31 - 1)}{-0.1 \times (15.3)^{10}(15.3 - 1)} = 0.663.10^{-17}$$

$$c_F^* = mq_F = 6.7 \times 150 = 1005 \text{ kg.m}^{-3}$$

$$p_3 = \frac{1005}{-0.663.10^{17}(15.13^{11} + 0.1) + 0.31^{11} - 1.054}$$

$$p_3 = -953.5$$

$$p_2 = 0$$

$$p_1 = 953.5 \times (0 + 1.054) = 1005$$

Hence:

$$c_o^* = p_1 + p_2 + p_3 = 1005 - 953.5 = 51.5$$

$$q_o = \frac{51.5}{6.7} = 7.69 \text{ kg.m}^{-3}$$

If we accept a true density of the cake of 1, its content in terms of residual oil is:

$$\frac{7.69}{1000} = 0.0077 < 1\%$$

and the extraction yield is:

$$\frac{150-7.69}{150}=94.8\%$$

NOTE.–

The above calculation shows that, if $c_F = 0$, we can accept the simplified formula:

$$q_o^* = q_F^* \left[1 - \frac{1}{g_3}\right]$$

The extraction yield is then simply:

$$\frac{q_F - q_o^*}{q_F^*} = \frac{1}{g_3}$$

NOTE.–

It is easy to verify that, if we were to apply the expression of the perfect countercurrent with $N_E = 7.83$, the extraction yield would be practically equal to 1, which is far greater than is the case in reality.

EXAMPLE 4.4 (Bagasse).–

Bagasse contains 15% ligneous material, and therefore 85% aqueous solutions:

$$m = \frac{1}{0.85} = 1.18 \varepsilon_h = 0.93 n = 9$$

The cane contained 15% sugar, and the yield of the mills was 70%. Thus, from the 15% sugar initially present in the cane, 4.5% remains.

$$\frac{Q_E}{Q_R} = 1 \quad \text{and} \quad \frac{mQ_E}{Q_R} = E = 1 \times 1.18 = 1.18$$

The radius r of the sugar molecule is such that:

$$\frac{4\pi r^3}{3} = \frac{362}{1590} \times \frac{1}{6.02.10^{26}}$$

362 is the molar mass of sugar and 1590 its density in $kg.m^{-3}$.

Hence:

$$r = 4.6.10^{-10}\ \text{m}$$

For the viscosity of the juice, we accept the mean value of 1.3×10^{-3} Pa.s. Let us use the Stokes–Einstein formula (equation [4.4]):

$$D = \frac{1.38.10^{-23} \times 350}{6\pi \times 1.3.10^{-3} \times 4.6.10^{-10}} = 4.28.10^{-10}\,\text{m}^2.\text{s}^{-1}$$

The diffusivity in the fibers is:

$$4.28.10^{-10} \times \frac{0.85}{2} = 1.82.10^{-10}\ \text{m}^2.\text{s}^{-1}$$

The passing of the cane through the hammer mill before pressing reveals fibers whose mean diameter is 0.8mm. The radius of the equivalent sphere for diffusion is:

$$\frac{0.8}{2}.10^{-3} \times 1.3 = 0.5.10^{-3}\text{m} \qquad [4.2]$$

Hence:

$$\frac{1}{K_c a_L} = \frac{1}{(1-0.93)} \times \frac{\left(0.5.10^{-3}\right)^2}{15 \times 1.82.10^{-10}} = 1308\ \text{s}$$

The device contains nine sections of 2.1 m, and the product advances at the rate of $0.7\ \text{m.mn}^{-1}$.

$$Q_R = (1 - 0.93)A \times \frac{0.7}{60} = (8.17.10^{-4} \times A)\ \text{m}^3.\text{s}^{-1}$$

$$\Omega_p = 9 \times 2.1 \times A = (18.9 \times A)\ \text{m}^3$$

$$\frac{\Omega_p}{Q_R} = \frac{18.9}{8.17.10^{-4}} = 23133\ \text{s}$$

$$\varepsilon N_E = \frac{23133 \times 1.18}{1308} = 20.86$$

We suppose that the device operates in the irrigated regime with a saturation of 0.1:

$$E = 1.18 \times 0.1 \times \frac{0.93}{1-0.93} = 1.57$$

$$\alpha = \left[\exp\left[\frac{20.86}{9}\right] - 1\right]^{-1} = 0.11$$

$$T_1 = -0.173 \qquad T_3 = -4.18$$

$$T_2 = 3.05 \qquad T_4 = 1.30$$

$$-0.173\,\lambda^2 + 2.88\lambda - 1.3 = 0$$

$$\lambda_2 = 16.18 \qquad g_2 = 1 - 0.11(15.18) = -0.67$$

$$\lambda_3 = 0.464 \qquad g_3 = 1 - 0.11(0.464 - 1) = 1.058$$

The extraction yield is:

$$\eta = \frac{1}{g_3} = \frac{1}{1.058} = 0.945$$

However, this value is probably optimistic.

4.4. Diffusion batteries

4.4.1. *Principle*

A diffusion battery consists of a series of columns traversed in sequence by the extract. After a period of time τ_c called the cycle time, the column having contained the product for the longest time is discharged and refilled with fresh product. Thus, it becomes the first column in relation to the product. If there are n columns, the time for which the product remains in a column is $n\tau_c$.

The device of a diffusion battery allows us to "circulate" the product and the extract in a countercurrent flow, although in reality, the product does not move.

Diffusion batteries have long been used to extract sugar from sugar beet, but nowadays, continuous devices are used for that purpose instead. However, instant coffee is still produced in this manner because, after extraction, the grains are subjected to hydrolysis under pressure and at 170°C, which would not be possible with a continuous device. In addition, as the contents of the column are under pressure, it is easy to drain it.

4.4.2. *Spaninks and Bruin's method [SPA 79a]*

Let x represent the distance counted in the direction of displacement of the liquid. In a slice of thickness Δx, we can write the balances:

$$A\Delta x\varepsilon\frac{\partial c}{\partial \tau}+Q_E\Delta c=K_c a_L\left(c^*-c\right)A\Delta x$$

$$A\Delta x\left(1-\varepsilon\right)\frac{\partial q}{\partial \tau}=-K_c a_L\left(c^*-c\right)A\Delta x$$

ε : porosity between the grains;

Q_E : flowrate of extract: $m^3.s^{-1}$;

A : section area of the column: m^2;

c : concentration of the extract: $kg.m^{-3}$;

K_c : transfer coefficient: $m.s^{-1}$;

a_L : surface area of transfer per unit volume of the bed: m^{-1};

q : concentration of soluble in the grains: $kg.m^{-3}$;

τ : time.

The above equations can be written:

$$\varepsilon\frac{\partial c}{\partial \tau}+\frac{Q_E}{A}\frac{\partial c}{\partial x}=K_c a_L\left(c^*-c\right)$$

$$\left(1-\varepsilon\right)\frac{\partial c}{\partial x}=-K_c a_L\left(c^*-c\right)$$

Over the duration of the extraction, the volume of solvent introduced is:

$$n\,\tau_c Q_E$$

where τ_c is the cycle time.

The volume of extract recovered is such that:

$$n\tau_c Q_E^* = n\tau_c Q_E - \varepsilon\Omega_u$$

Ω_u is the volume of a column.

Indeed, at the rich end, the extract is used for the soaking and swelling of the product introduced.

The extraction factor is then defined by:

$$\mathcal{E} = m\frac{Q_E^*}{Q_R} = m\left[\frac{Q_E - \varepsilon\Omega_u / n\tau_c}{(1-\varepsilon)\Omega_u / \tau_c}\right]$$

m is the equilibrium coefficient:

$$c = mq^*$$

The entrainment parameter E is the ratio of the volume of liquid entrained by the discharged solid to the volume of solid, all multiplied by m:

$$E = m\frac{\varepsilon\Omega_u}{(1-\varepsilon)\Omega_u} = \frac{m\varepsilon}{1-\varepsilon}$$

The practical number of transfer units on the side of the extract is defined by:

$$N_{EP} = \frac{K_c a_L \Omega_p}{Q_E - \varepsilon\Omega_u / n\tau_c} \qquad [4.11]$$

where:

$$\Omega_p = n\Omega_u \quad (m^3)$$

n is the number of extraction columns.

The ideal number of transfer units is then given by:

$$N_{EI} = N_{EP}\left[1 - 0.59\, n^{-1.17} \times \mathcal{E}^{-0.3} \times N_{EP}^{0.15\mathcal{E}n} \times E^{0.33}\right] \qquad [4.12]$$

EXAMPLE 4.5 (Instant coffee).–

Consider an array for the extraction of instant coffee, made up of 6 columns:

1 column is assigned to draining and filling (with soaking by rich liquor and swelling of the grains).

1 column is assigned to the hydrolysis of the product at 170°.

We have four remaining columns where extraction, in the true sense of the word, takes place. The columns are 5 m tall and 1 m in diameter. The unit volume of a column is:

$$\Omega_u = \frac{\pi}{4} \times 1^2 \times 5 = 3.93 \text{m}^3$$

In beds of wash, the porosity accessible to the liquor is 0.3. The cycle time τ_c is half an hour, which is 1800 seconds. The "flowrate" of product is:

$$Q_R = (1-\varepsilon) \times \frac{\Omega_u}{\tau_c} = 0.7 \times \frac{3.39}{1800} = 1.53.10^{-3} \text{m}^3.\text{s}^{-1}$$

For 100 kg of product, we add 240 kg of water enriched with hydrolyable species (20 kg of hydrolyzables).

Dry-roasted ground coffee has a total porosity equal to 0.5, and its true density is 660 kg.m^{-3}. The apparent volume occupied by 100 kg of product is:

$$\frac{100}{660} \times \frac{1}{(1-0.5)} = 0.303 \text{ m}^3$$

The 100 kg of product are made up as follows:

Insoluble solid:	55
Soluble:	25
Hydrolyzables:	20

After soaking and swelling, the porosity drops to 0.3 and the volume occupied by the grains is:

$$0.303 \times (1 - 0.3) = 0.212 \text{ m}^3$$

To treat 100 kg of product, we inject the following at the poor end of the installation:

240 kg	of water
20 kg	of hydrolyates

The flowrate of that liquor is:

$$Q_E = Q_R \left[\frac{240/1000 + 20/660}{0,212} \right] = 1,95.10^{-3} m^3.s^{-1}$$

After swelling of 100 kg of product, the volume of the solid matrix is:

Insoluble solid	55/660	=	0.083	
Hydrolyzable	20/660	=	0.030	
Swell water	25/1000	=	0.025	
		Total	0.138	m^3

The accessible porosity (which allows the water to enter and dissolve the soluble in the grains) is:

$$\varepsilon_A = \frac{0.212 - 0.138}{0.212} = 0.35$$

The equilibrium coefficient is then:

$$m = \frac{1}{\varepsilon_A} = \frac{1}{0.35} = 2.86$$

The extraction factor is therefore:

$$\varepsilon = 2.86 \left[\frac{1.95.10^{-3} - 0.3 \times 3.93/(1800 \times 4)}{3.93(1-0.3)/1800} \right] = 3.04$$

Indeed, a portion of the liquid is used to fill the interstitial voids of the wash.

The factor E is:

$$E = 2.86 \times \frac{0.3}{1-0.3} = 1.23$$

The phenomenon entering into play here is not the mutual diffusion of two miscible liquids, but rather the (significantly slower) dissolution of the amorphous soluble in the water.

Here, the following value will be adopted:

$$K_c a_L = 5.64.10^{-4}(1-\varepsilon)\ (s^{-1})$$

Indeed, this value cannot be calculated. Finally:

$$K_c a_L = 5.64.10^{-4}(1-0.3) = 3.93.10^{-4} s^{-1}$$

The practical number of transfer units is (see equation [4.11]):

$$N_{EP} = \frac{3.93.10^{-4} \times 3.93 \times 4}{1.95.10^{-3} - \dfrac{0.3 \times 3.93}{4 \times 1800}} = 3.46$$

Hence, the ideal number of transfer units is (see equation [4.12]):

$$N_{EI} = 3.46[1 - 0.59 \times 4^{-1.17} \times 3.04^{-0.3} \times 3.46^{0.15 \times 3.04} \times 1.23^{0.33}]$$

$$N_{EI} = 3.1$$

The ideal extraction factor is:

$$\varepsilon_I = 2.86 \times \frac{1.95.10^{-3}}{1.53.10^{-3}} = 3.65$$

$$\exp[(\varepsilon_I - 1)N_{EI}] = \exp(2.65 \times 3.1) = 3695$$

However, the concentration of the incoming product is:

$$q_e = \frac{25}{0.212} = 118\ kg.m^{-3}$$

The concentration of the incoming extract is:

$$c_e = \frac{20}{\dfrac{240}{1000} + \dfrac{20}{660}} = 74\ kg.m^{-3}$$

Remember that:

$$m = 2.86$$

Hence:

$$3695 = \frac{2.86 \times 118 - c_s}{82 \times 2.65 + 2.86 \times 118 - 3.65 c_s}$$

The concentration of the liquor at the output is then:

$$c_s = 142 \text{ kg.m}^{-3}$$

The concentration of the wash at output is given by the balance:

$$1.95.10^{-3}(142 - 74) = 1.53.10^{-3}(118 - q_s)$$

$$q_s = 31 \text{ kg.m}^{-3}$$

4.5. Washing of ores

4.5.1. *Concentrations and gravimetric fractions*

The washing of ores is often performed in a series of stages, each composed of a mixing tank and a gravity decanter. We can graphically calculate the number of stages, using the gravimetric fractions.

Let us first examine the correspondence between the gravimetric fractions and the concentrations. We use the indices I, A and U, respectively, for the insoluble solid, the solvent and the soluble.

1) In the liquor

$$Y_A + Y_U = 1 \quad [4.13]$$

Y_A and Y_U are mass fractions.

If ρ_A and ρ_U are the densities of the solvent and of the soluble, the volume occupied by 1 kg of liquor is:

$$V_L = \frac{Y_A}{\rho_A} + \frac{Y_U}{\rho_U} \quad [4.14]$$

The concentration in terms of soluble is:

$$c = \frac{Y_U}{V_L} \quad [4.15]$$

Equations [4.14] and [4.15], by elimination of V_L, give us the following:

$$Y_A = \rho_A Y_U \left[\frac{1}{c} - \frac{1}{\rho_U}\right]$$

and with equation [4.13]:

$$Y_U = \frac{c\rho_U}{c(\rho_U - \rho_A) + \rho_U\rho_A} \text{ et } Y_A = \frac{\rho_U\rho_A - c\rho_A}{c(\rho_U - \rho_A) + \rho_U\rho_A}$$

2) In the particles

Below represents the mass fractions.

$$X_I + X_A + X_U = 1 \quad [4.16]$$

$$V_p = \frac{X_I}{\rho_I} + \frac{X_A}{\rho_A} + \frac{X_U}{\rho_U} \quad [4.17]$$

$$q = \frac{X_U}{V_p} \quad [4.18]$$

An additional equation is obtained by writing that, in 1 kg of particles, the volume of the accessible particle pores occupied by the liquid is:

$$\varepsilon_A V_p = \frac{X_A}{\rho_A} + \frac{X_U}{\rho_U} \quad [4.19]$$

By eliminating V_p between equations [4.18] and [4.19], we obtain:

$$X_A = X_U\rho_A\left[\frac{\varepsilon_A}{q} - \frac{1}{\rho_U}\right] \quad [4.20]$$

When we eliminate V_p between equations [4.17] and [4.18], we obtain:

$$\frac{X_U}{q} = \frac{X_I}{\rho_I} + \frac{X_A}{\rho_A} + \frac{X_U}{\rho_U} \quad [4.21]$$

According to equations [4.16] and [4.20]:

$$X_I = 1 - X_A - X_U = 1 - X_U\left[1 + \frac{\varepsilon_A\rho_A}{q} - \frac{\rho_A}{\rho_U}\right] \quad [4.22]$$

Into equation [4.21], let us feed the expression of X_A given by equation [4.20] and the expression of X_I given by equation [4.22]. We obtain:

$$X_U = \frac{q}{D}$$

$$X_A = \frac{\varepsilon_A\rho_A - q\frac{\rho_A}{\rho_U}}{D}$$

$$X_I = \frac{\rho_I(1-\varepsilon_A)}{D}$$

where:

$$D = \rho_I + q + \rho_A\varepsilon_A - q\frac{\rho_A}{\rho_U} - \varepsilon_A\rho_I$$

3) Equilibrium coefficient

We know that the equilibrium coefficient is defined (in terms of concentrations) by:

$$m = \frac{1}{\varepsilon_A} = \frac{c}{q} = \frac{Y_U V_P}{X_U V_L}$$

From this relation, we derive the sharing coefficient for the gravimetric fractions:

$$\frac{Y_U}{X_I} = \frac{V_L}{V_p\varepsilon_A}$$

where, it must be remembered, V_L and V_p are respectively the volumes of 1 kg of liquor and 1 kg of impregnated particles. These volumes may vary slightly with the composition of the liquor. Thus, in terms of gravimetric fractions, the equilibrium coefficient is generally not constant.

4.5.2. *Graphical determination of a number of stages of washing*

The method is identical to that used for liquid–liquid extraction (see Chapters 1 and 2). Only the following details are different:

– in the washing of ore, we never use the molar fractions, working instead with the gravimetric fractions. However, everything that has been said about barycentric coordinates and the triangular representation remains valid with the gravimetric fractions;

– the typical representation uses not an equilateral triangle, but a right-angled triangle. Here again, though, the properties of the representation in barycentric coordinates remain valid.

As indicated in Figure 4.4, the vertex A corresponds to the solvent, the vertex U to the soluble and the vertex I to the insoluble. The side A U represents the compositions of the wash liquor (which can only be made up of differing fractions of insolubles).

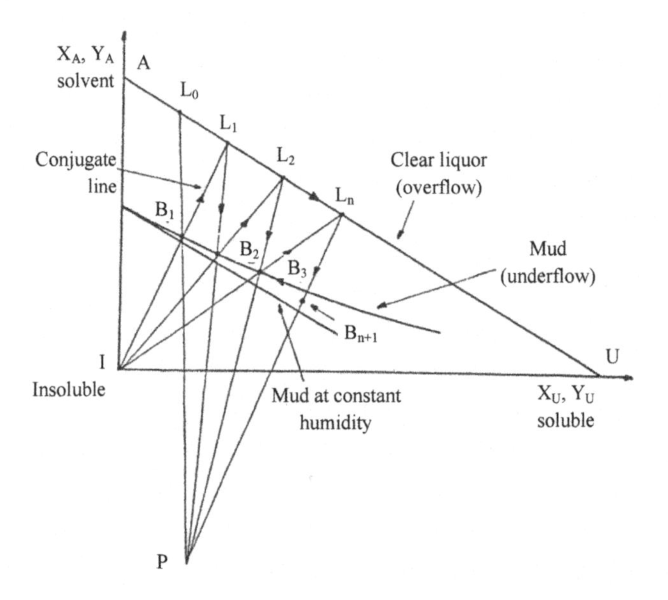

Figure 4.4. *Triangular representation*

In the particles, the gravimetric fraction of insolubles is almost constant. Indeed, the mass of insolubles is constant but the mass of the particles may vary slightly if the composition (and therefore the density) of the liquor filling the pores varies. Finally, the curve representative of the compositions of particles at equilibrium with the liquid is near to a straight line parallel to the side A U of the triangle. The true curve strays somewhat from that straight line at contents high in solubles, because the soluble is generally two-to-three times denser than water (the typical solvent), which increases the mass of the pores and consequently decreases the gravimetric fraction of insoluble.

The conjugate lines expressing the equilibrium between the mud B and the clear liquor L all pass through the point I. Indeed, at equilibrium, the composition of the external liquor is identical to the composition of the liquor filling the pores. For the solvent and the soluble, we have:

$$\varepsilon_A c_A = q_A$$

$$\varepsilon_A c_U = q_U$$

That is:

$$\frac{c_A}{c_U} = \frac{q_A}{q_U} \qquad \frac{Y_A}{Y_U} = \frac{X_A}{X_U}$$

Hence, the slope of the line joining the point I to the image of the solution is identical to that of the line joining I to the image of the particles; thus, the three points are aligned.

In reality, it must not be believed that the liquor circulates in one direction and the solid particles circulate in the other direction with no liquor to accompany them. Gravity settlers deliver mud, rather than a damp divided solid. However, if the volumetric fraction of liquor in the mud is constant throughout the installation, the above discussion remains valid, on condition that we include in the accessible porosity ε_A the volume of the liquor which accompanies the particles in the mud. Thus, if the liquors are rich in solubles, they will be denser, the particles will settle less quickly and the mud will be more dilute, which decreases the gravimetric fraction X_I of insolubles and accounts for the fact that the curve of the images of muds shifts toward the solubles and moves away from the ideal line as Y_A decreases.

Toward the end of the installation that is rich in soluble, two factors are at work:

– the soluble replaces the solvent in the mud, and therefore X_A decreases progressively;

– however, as we have seen, the mud is more dilute, which decreases the extent of the above effect.

Finally, if we take $y = X_A/X_I$ as a function of $x = X_U/X_I$, we obtain a curve having the shape in Figure 4.5. This curve is none other than the characteristic curve of the real muds in Figure 4.4, but expressed differently.

This curve can be exploited as follows.

Suppose we feed the installation with a mud whose composition is such that the partial flowrates at the inlet of insoluble, solvent and soluble are W_I, W_A and W_U. Furthermore, we impose a purification yield η for the installation. At the output, the flowrates are:

Insoluble $\quad W_I^o = W_I^i$

Soluble $\quad W_U^o = (1 - \eta)W_U^i$

The exponents i and o refer respectively to the inlet and the outlet.

We calculate $x = W_U/W_I$ and the curve in Figure 4.5 gives us $y = W_A/W_I$. From this we deduce $W_A = yW_I$. Knowing the three flowrates W_I, W_U and W_A at output, it is easy to deduce the composition of the muds at that point.

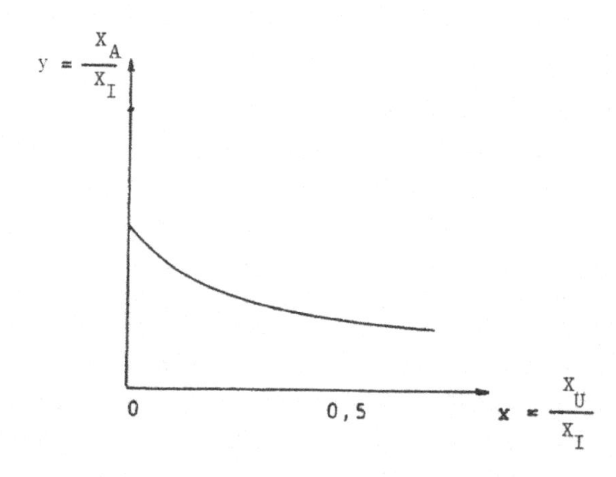

Figure 4.5. *Richness of the mud in terms of solvent and soluble*

In addition, if we have chosen the flowrate and the composition of the wash liquor entering into the installation, the balances of solvent and soluble can be used to deduce the composition of the output liquor.

Finally, in Figure 4.4, we can place the following four points:

L_n : outlet wash liquor;

B_{n+1} : input mud;

L_0 : input wash liquor;

B_1 : outlet washed mud.

The intersection of the lines $L_n B_{n+1}$ and L_0B_1 gives the pole P of the construction. The operating lines will pass through that pole. We then begin with the end poor in soluble (index zero), alternating the conjugate lines and the operating lines. We stop when we reach or pass the line $L_n B_{n+1}$. The number of stages needed is equal to the number of conjugate lines that we have plotted.

Appendix

Characteristics of Typical Packings

Material	Dimension (mm)	Thickness of the wall (mm)	Apparent density ($kg.m^{-3}$)	Volumetric surface (m^{-1})	No-load fraction (%) (Porosity)
Ceramic Raschig rings	102	11	600	50	75
	76	9.5	650	70	75
	51	6.5	650	95	75
	38	6.5	700	130	68
	25	3	700	200	73
	19	2.5	700	240	72
	13	2.5	800	370	64
	9.5	1.5	800	500	65
	8	1	800	600	72
	6.5	0.8	800	800	70
Steel Raschig rings	76	1.6	450	70	94
	51	1.2	460	100	94
	38	0.9	480	140	94
	25	0.7	560	210	93
	19	0.6	580	270	93
	13	0.5	700	400	91
	9.5	0.5	930	600	88
	6.3	0.5	1 400	800	82

Ceramic Pall rings	102	9.5	420	56	82
	51	5	550	125	78
	25	3	640	220	73
Steel Pall rings	51	1	400	105	95
	35	0.8	430	145	95
	25	0.6	500	240	94
	16	0.4	550	370	93
Porcelain Berl saddles	51		640	110	77
	38		610	150	76
	25		720	250	70
	19		800	300	67
	13		900	480	65
	6.3		900	1 000	62
Intalox saddles	51		600	110	75
	38		600	160	74
	25		600	250	75
	19		600	300	73
	13		600	480	73

Bibliography

[AGA 76] AGARWAL J.C., KLUMPAR I.V., "Multistage-leaching simulation", *Chemical Engineering*, pp. 135–140, May 1976.

[AGU 84] AGUERRE R.J., SUAREZ C., VIOLLAZ P.E., "Calculation of the variation of the heat of desorption with moisture content on the basis of the BET Theory", *J. of Food Technology*, vol. 19, p. 325, 1984.

[AGU 85] AGUERRE R.J., GABITTO J.F., CHIRILE J., "Utilisation of Fick's second law for the evaluation of diffusion coefficients in food processes controlled by internal diffusion", *Journal of Food Technology*, vol. 20, pp. 623–629, 1985.

[ATW 84] ATWOOD J.G., GOLDSTEIN J., "Measurements of diffusion coefficients in liquids at atmospheric and elevated pressure by the chromatographie broadening technique", *J. Phys. Chem.*, vol. 88, p. 1875, 1984.

[BOE 85] BOEDEKER E., "Ueber das Verhältnis zwischen Masse und Wirkung beim Kontact ammoniakalischer Flüssigkeiten mit Ackererde und mit kohlensaurem Kalk", *Journal für Landwirtschaft*, vol. 7, pp. 48–58, 1885.

[BRU 38] BRUNAUER S., EMMETT P.H., TELLER E., "Adsorption of gases in multimolecular layers", *J. of the Am. Chem. Society*, vol. 60, p. 309, 1938.

[BRU 40] BRUNAUER S., DEMING L.S., DEMING W.E. *et al.*, "On a theory of the van der Walls adsorption of gases", *J. of the Am. Chem. Soc.*, vol. 62, p. 1723, 1940.

[CAM 57] CAMBRONY H.R., "Les techniques de fabrication du café soluble", *Café, Cacao*, vol. 1, no. 2, pp. 65–74, May-August 1957.

[CHE 70] CHEN N.H., "Optimum theoretical stages in countercurrent leaching", *Chemical Engineering*, pp. 71–74, 24 August 1970.

[COU 79] COULSON J.M., RICHARDSON J.F., *Chemical Engineering*, Vol. III, Second edition, Pergamon, 1979.

[CRA 56] CRANK J., *The Mathematics of Diffusion*, Oxford University Press, London, 1956.

[CZO 90] CZOK M., GUIOCHON G., "The physical sense of simulation models of liquid chromatography: propagation through a grid or solution of the mass balance equation?", *Anal. Chem.*, vol. 62, p. 189, 1990.

[DAN 80a] DANESI P.R., CHIARIZIA R., "The kinetics of metal solvent extractions", *C.R.C. Critical Reviews in Analytical Chemistry*, vol. 10, no. 1, pp. 1–125, 1980.

[DAN 80b] DANESI P.R., CHIARIZIA R., VAN DE GRIFT G.F., "Kinetics and mechanism of the complex formation reactions between Cu(II) and Fe(III) aqueous species and a β-Hydroxy Oxime in toluene", *Journal of Physical Chemistry*, vol. 84, no. 25, pp. 3455–3461, 1980.

[DAN 80c] DANESI P.R., VAN DE GRIFT G.P., HORWITZ E.P. *et al.*, "Simulation of interfacial two-step consecutive reactions by diffusion in the mass-transfer kinetics of liquid-liquid extraction of metal cations", *Journal of Physical Chemistry*, vol. 84, no. 26, pp. 3582–3587, 1980.

[DAN 84] DANESI P.R., "The relative importance of diffusion and chemical reactions in liquid-liquid extraction kinetics", *Solvent Extractions and Ion Exchange*, vol. 2, no. 1, pp. 29–44, 1984.

[DED 67] DEDRICK R.L., BECKMANN R.B., "Kinetics of adsorption by activated carbon from dilute aqueous solution", *Chem. Eng. Progr. Symp. Series*, vol. 63, p. 68, 1967.

[DUL 70] DULLIEN F.A.L., BATRA V.K., "Determination of the structure of porous media", *Ind. Eng. Chem.*, vol. 62, p. 25, 1970.

[DUR 16a] DUROUDIER J.-P., *Thermodynamics*, ISTE Press, London and Elsevier, Oxford, 2016.

[DUR 16b] DUROUDIER J.-P., *Adsorption-Dryers for Divided Solids*, ISTE Press, London and Elsevier, Oxford, 2016.

[DUR 16c] DUROUDIER J.-P., *Liquid-Solid Separators*, ISTE Press, London and Elsevier, Oxford, 2016.

[DUR 16d] DUROUDIER J.-P., *Crystallization and Crystallizers*, ISTE Press, London and Elsevier, Oxford, 2016.

[DYE 66] DYER D.F., SUNDERLAND J.E., "Bulk and diffusionnal transport in the region between molecular and viscous flow", *J. Heat Mass Transfer*, vol. 9, p. 519, 1966.

[EL 77] EL SABAWI M., PEI D.C.T., "Moisture isotherms of hygroscopie porous solids", *Ind. Eng. Chem. Fund*, vol. 16, p. 321, 1977.

[FRE 94] FREY D.D., RODRIGUES A.E., "Explicit calculation of multicomponent equilibria for ideal adsorbed solutions", *A.I. Ch. E. Journal*, vol. 40, p. 182, 1994.

[FRI 74] FRITZ W., SCHLUENDER E.U., "Simultaneous adsorption equilibria of organic solutes in dilute aqueous solutions on activated carbon", *Chem. Eng. Science*, vol. 29, p. 1279, 1974.

[FUR 76] FURUSAWA T., SUSUKI M., SMITH J.M., "Rate parameters in heterogeneous catalysis by pulse techniques", *Catalysis Review*, vol. 13, p. 43, 1976.

[GAR 72] GARG D.R. RUTHVEN D.M., "The effect of the concentration dependence of diffusivity on zeolitic sorption curves", *Chem. Eng. Science*, vol. 27, p. 417, 1972.

[GAY 53] GAYLER R., ROBERTS N.W., PRATT H.R.C., "Liquid-liquid extraction, Part IV, A further study of holdup in packed columns", *Trans. Inst. Chem. Eng.*, vol. 31, pp. 57–68, 1953.

[GIL 58] GILLILAND E.R. BADDOUR R.F., RUSSEL J.L., "Rate of flow through microporous solids", *A.I. Ch. E Journal*, vol. 4, p. 90, 1958.

[GLU 55a] GLUECKAUF E., "Formula for diffusion into spheres and their application to chromatography", *Trans. Faraday Soc*, vol. 51, p. 1540, 1955.

[GLU 55b] GLUECKAUF E., "Theory of chromatography-Part 9 – The theoretical phase concept in column separations" *Trans. Faraday Soc*, vol. 51, p. 34, 1955.

[GOL 89a] GOLDHAN-SHIRAZI S., GUICHON G., "Analytical solution for the ideal model of chromatography in the case of a pulse of a binary mixture with competitive Langmuir esotherm", *J. Phys. Chem.*, vol. 93, p. 4143, 1989.

[GOL 89b] GOLDHAN-SHIRAZI S., GUICHON G., "Theory of optimization of the experimental conditions of preparative elution using the ideal model of liquid chromatography", *Nal. Chem.*, vol. 61, p. 1276, 1989.

[GRI 03] GRITTI F., GUIOCHON G., "New Thermodynamically consistent competitive adsorption isotherm in RPLC", *J. of Colloid and Interface Science*, vol. 264, pp. 43–59, 2003.

[HAL 66] HALL K.R., EAGLETON L.C., ACRIVOS A. *et al.*, "Pore and solid diffusion kinetics in fixed-bed adsorption under constant pattern conditions", *I.E.C. Fund*, vol. 5, p. 212, 1966.

[HAS 76] HASHIMOTO K., MIURA K., "A simplified method to design fixed-bed adsorbers for the Freundlich isotherm", *J. of Chem. Eng. Japan*, vol. 9, p. 388, 1976.

[HAY 73] HAYNES H.W. JR., SARMA P.N., "A model for the application of gas chromatography to measurements of diffusion in biodisperse structured catalysts", *A.I. Ch.E. Journal*, vol. 19, no. 5, pp. 1043–1046, 1973.

[HOO 67] HOORY S.E., PRAUSNITZ J.M., "Molecular thermodynamics of monolayers gas adsorption in homogeneous and heterogeneous solid surfaces", *Chem. Eng. Progr. Symp. Series*, vol. 63, p. 3, 1967.

[HOR 78] HORVATH C., LIN H.-J., "General plate height equation and a method for the evaluation of the individual plate height contributions", *J. of Chromatrography*, vol. 149, p. 43, 1978.

[HUA 73] HUANG T.-C., LI K-Y., "Ion exchange kinetics for calcium radiotracer in a batch system", *Ind. Eng. Chem. Fund.*, vol. 12, p. 50, 1973.

[JUR 77] JURY S.H., "Diffusion in tabletted catalysts', *The Canadian Journal of Chemical Engineering*, vol. 55, p. 538, 1977.

[JUR 78] JURY S.H., "Transport phenomena in heterogeneous media", *J. of the Franklin Institute*, vol. 305, p. 79, 1978.

[KAL 97] KALAICHELVI P., MURUGESAN T., "Prediction of slip velocity in rotating disc contactors", *J. Chem. Tech. Biotechnol.*, vol. 69, pp. 130–136, 1997.

[KIM 87] KIM C.J., BUYN S.M., CHEIGH H.S. *et al.*, "Comparison of solvent extraction characteristics of rice bran pretreated by hot air drying, steam cooking and extrusion", *J.A.O.C.S.*, vol. 64, no. 4, pp. 514–516, 1987.

[KJA 73] KJAERGAARD O.G., ANDRESON E., "Preparation of coffee extracts by continuous extraction", *ASIC 6th Annual Conference-Bogota*, pp. 234–239, 1973.

[KNO 77] KNOX J.H., "Practical aspects of L.C. theory", *J. of Chromatography*, vol. 15, p. 352, 1977.

[KUM 85] KUMAR A., HARTLAND S., "Gravity settling in liquid-liquid despersions", *The Canad. Journal of Chem. Eng.*, vol. 63, pp. 368–376, 1985.

[KUM 86] KUMAR A., HARTLAND S., "Prediction of drop size in rotating disc extractor", *The Canad. Journal of Chem. Eng.*, vol. 64, pp. 915–924, 1986.

[KUM 87] KUMAR A., HARTLAND S., "Prediction of dispersed phase holdup in rotating disc contactor", *Chem. Commun.*, vol. 56, pp. 87–106, 1987.

[KUM 88a] KUMAR A., HARTLAND S., "Prediction of dispersed phase holdup and flooding velocities in Karr reciprocating plate extraction columns", *Ind. Eng. Chem. Res.*, vol. 27, pp. 131–138, 1988.

[KUM 88b] KUMAR A., HARTLAND S., "Prediction of dispersed phase holdup in pulsed perforated plate extraction columns", *Chem. Eng. Process*, vol. 23, pp. 41–59, 1988.

[KUM 89a] KUMAR A., HARTLAND S., "Independent prediction of slip velocity and holdup in liquid-liquid extraction columns", *The Canad. Journal of Chem. Eng.*, vol. 67, pp. 17–25, 1989.

[KUM 89b] KUMAR A., HARTLAND S., "Prediction of continuous-phase axial mixing coefficients in pulsed perforated-plate extraction columns", *Ind. Eng. Chem. Res.*, vol. 28, pp. 1507–1513, September 1989.

[KUM 92] KUMAR A., HARTLAND S., "Predicting of axial mixing coefficient in rotating disc and asymetric rotating disc extraction columns", *The Canad. Journal of Chem. Eng.*, vol. 70, pp. 77–87, 1992.

[KUM 94] KUMAR A., HARTLAND S., "Prediction of drop size dispersed phase holdup, slip velocity and limiting throughputs in packed extraction columns", *Trans. Inst. Chem. Eng.*, vol. 72, Part A, January 1994.

[KUM 95] KUMAR A., HARTLAND S., "A unified correlation for the prediction of dispersed phase holdup in liquid-liquid extraction columns", *Ind. Eng. Chem. Res.*, vol. 34, pp. 3925–3940, 1995.

[KUM 96] KUMAR A., HARTLAND S., "Unified correlations for the prediction of drop size in liquid-liquid extraction columns", *Ind. Eng. Chem. Res.*, vol. 35, pp. 2682–2695, 1996.

[KUM 99a] KUMAR A., HARLAND S., "Computational strategies for sizing liquid-liquid extractors", *Ind. Eng. Chem. Res.*, vol. 38, pp. 1040–1056, 1999.

[KUM 99b] KUMAR A., HARTLAND S., "Correlations for prediction of mass transfer coefficients in single drop systems and liquid-liquid extraction columns", *Trans. Instr. Chem. Eng.* 77, Part A, pp. 372–384, 1999.

[LAD 78a] LADDHA G.S., DEGALEESAN T.E., KANNAPPAN R., "Hydrodynamics and mass transport in rotary disc contactors", *The Canadian Journal of Chem. Eng.*, p. 56, April 1978.

[LAD 78b] LADDHA G.S., DEGALEESAN T.E., *Transport Phenomena in Liquid-Liquid Extraction*, McGraw Hill, New York 1978.

[LAN 16] LANGMUIR I., "The constitution and fundamental properties of solids and liquids, Part I, Solids", *J. of Amer. Chem. Society*, vol. 38, p. 2221, 1916.

[LAN 18] LANGMUIR I., "The adsorption of gases on plane surfaces of glass, mica and platinium", *J. of Amer. Chem. Society*, vol. 40, p. 1361, 1918.

[LEE 86] LEE A.K.K., BULLEY N.R., FATTORI M. *et al.*, "Modelling of supercritical carbon dioxide extraction of canola oilseed in fixed beds", *J.A.O.C.S.*, vol. 63, no. 7, pp. 921–925, 1986.

[LEN 32] LENNARD-JONES J.E., "Processes of adsorption and diffusion on solid surfaces", *Trans. of the Faraday Society*, vol. 28, p. 333, 1932.

[LEW 50] LEWIS W.K., GILLILAND E.R., CHERTOW B. *et al.*, "Adsorption equilibria. Hydrocarbon gas mixtures", *I.E.C.*, vol. 42, p. 1319, 1950.

[LIN 89] LIN B., GUIOCHON G., "Numerical simulation of chromatographic band profiles at large concentrations: bength of space increment and height equivalent to a theorical plate", *Separation Science and Technology*, vol. 24, nos. 1 & 2, p. 31, 1989.

[MAR 87] MARTIRE D.E., "Unified Theory of adsorption chromatography: gas, liquid and supercritical fluid mobile phases", *J. of Liquid Chromatography*, vol. 10, p. 1569, 1987.

[MAR 88] MARTIRE D.E., "Unified Theory of adsorption chromatography: gas, liquid and supercritical fluid mobile phase", *J. of Chromatography*, vol. 452, p. 17, 1988.

[MIC 52] MICHAELS A.S., "Simplified method of interpreting kinetic data in fixed-bed ion exchange", *Ind. Eng. Chem.*, vol. 44, p. 1922, 1952.

[MIL 76] MILLIGAN E.D., "Survey of current solvent extraction equipment", *Journal of American Chemist's Society*, vol. 53, pp. 286–290, June 1976.

[MIY 01] MIYABE K., GUIOCHON G., "Correlation between surface diffusion and molecular diffusion in reversed phase liquid chromatography", *J. Phys. Chem.*, vol. 105, p. 9202, 2001.

[MOR 91] MOREAU M., VALENTIN P., VID-MADJAR C. *et al.*, "Adsorption isotherm model for multicomponent adsorbate-adsorbate interactions", *J. of Colloid and Interface Science*, vol. 141, p. 127, 1991.

[MYE 65] MYERS A.L., PRAUSNITZ J.M., "Thermodynamics of mixed-gas adsorption", *A.I. Ch. E. Journal*, vol. 11, p. 121, 1965.

[MYE 83] MYERS A.L., "Activity coefficients of mixtures adsorbed on heterogeneous surfaces", *A.I. Ch. E. Journal*, vol. 29, p. 691, 1983.

[NGU 78] NGUYEN H.X., "Calculating actual stages in countercurrent equipment", *Chemical Engineering*, pp. 121–122, 6 November 1978.

[PID 35] PIDGEON L.M., "The sorption of vapors by active silica", *Canadian Journal of Research*, vol. 12, pp. 41–56, 1935.

[RAD 72] RADKE C.J., PRAUSNITZ J.M., "Thermodynamics of multi-solute adsorption from dilute liquid solutions", *A.I. Ch. E. Journal*, vol. 18, p. 761, 1972.

[REI 76] REIN P.W., "Extraction performance of a diffuser using a mathematical model", *The Sugar Journal*, pp. 15–23 December 1976.

[RIC 65] RICKLES R.N., "Liquid-solid extraction", *Chemical Engineering*, pp. 157–172, 15 March 1965.

[ROT 63] ROTHFELD L.B., "Gazeous counter diffusion in catalyst pellets", *A.I.Ch.E. J.*, vol. 9, p. 19, 1963.

[ROU 60] ROUNSLEY R.R., "Multimolecular adsorption equation", *A.I.Ch.E.*, vol. 7, p. 309, 1960.

[ROU 66] ROUND G.F., MABGOOD H.W., NEWTON R., "A numerical analysis of surface diffusion in a binary adsorbed film", *Separation Science*, vol. 1, nos. 2 & 3, p. 219, 1966.

[RUC 71] RUCKENSTEIN E., VAIDYANATHAN YOUNGQUIST G.R., "Sorption by solids with bidisperse pore structure", *Chem. Eng. Science*, vol. 26, p. 1305, 1971.

[RUT 71] RUTHVEN D.M., "Simple theoretical adsorption isotherm for zeolites", *Nature Physical Science*, vol. 222, p. 70, 1971.

[RUT 84] RUTHVEN D.M., *Principles of Adsorption and Sorption Processes*, John Wiley and Sons, New York 1984.

[SEE 82] SEEWALD H., "Charakterisierung von Mikroporen und ihre Bedeutung für die Kohletechnik", *Erdol und Kohle, Erdgas, Petrochemie*, vol. 35, p. 418, 1982.

[SEI 88] SEIBERT A.F., FAIR J.R., "Hydrodynamics and Man Transfer in Spray and Packed Liquid-Liquid Extraction Columns", *Ind. Eng. Chem. Res*, vol. 27, pp. 470–481, 1988.

[SIV 79] SIVETZ M., DESROSIER N.W., *Coffee Technology*, AVI Publishing Company Inc., Connecticut, 1979.

[SKE 79] SKELLAND A.H.P., HUANG Y.F., *A.I. Ch. E. J.*, vol. 25, p. 80, 1979.

[SMI 64] SMITH T.G., DRANOFF J., "Film diffusion controlled kinetics in binary ion exchange", *I.E.C. Fund.*, vol. 3, p. 195, 1964.

[SMO 59] SMOOT L.D., MAR B.W., BALB A.L., "Flooding characteristics and separation efficiencies of pulsed sieve plate extraction columns", *Ind. and. Eng. Chem.*, vol. 51, no. 9, pp. 1005–1010, September 1959.

[SPA 79a] SPANINKS J.A.M., BRUIN S., "Mathematical simulation of the performance of the solid-liquid extractors, Part I Diffusion Batteries", *Chemical Engineering Science*, vol. 34, pp. 199–205, 1979.

[SPA 79b] SPANINKS J.A.M., BRUIN S., "Mathematical simulation of the performance of the solid-liquid extractors, Part II Belt type extractors", *Chemical Engineering Science*, vol. 34, pp. 207–2015, 1979.

[THO 57] THORNTON J.D., "Liquid-liquid extraction, Part XIII: the effect of pulse wave-form and plate geometry on the performance and throughput of a pulsed column", *Trans. Instn. Chem. Engrs.*, vol. 35, pp. 316–330, 1957.

[TÓT 71] TÓTH J., "State equation of the solid-gas interface: layers", *Acta Chimica Academiae Scientiarium Hungaricae*, vol. 69, p. 311, 1971.

[TÓT 03] TÓTH J., "On thermodynamical inconsistency of isotherm equations: Gibbs thermodynamics", *J. of Colloid and Interface Science*, vol. 262, p. 25, 2003.

[TRE 63] TREYBAL R.E., *Liquid Extraction*, Mc Graw-Hill, 1963.

[UMP 01] UMPLEBY R.J., "Characterization of molecularly imprinted polymers with the Langmuir-Freundlich isotherm", *Analytical Chemistry*, vol. 73, p. 4584, 2001.

[VAL 76a] VALENTIN P., "Determination of gas-liquid and gas-solid equilibrium isotherms by chromatography: I. Theory of the step-and-pulse method", *Journal of Chromatographic Science*, vol. 14, p. 56, 1976.

[VAL 76b] VALENTIN P., "Determination of gas-liquid and gas-solid equilibrium isotherms by chromatography. II. Apparatus, specifications and results", *Journal of Chromatographic Science*, vol. 14, p. 132, 1976.

[VER 66] VERMEULEN T., MOON J.S., HENNICQ A. *et al.*, "Axial dispersion in extraction columns", *Chem. Eng. Progr.*, vol. 62, no. 9, 1966.

[VOL 25] VOLMER M., "Thermodynamische Folgerungen aus der Zustandsgleichung für adsorbierte Stoffe", *Zeitschrift für Physikalische Chemie*, vol. 115, p. 253, 1925.

[WAK 58] WAKAO N., OSHIMA T., YAGI S., "Mass transfer from particles to fluid in packed beds", *Kagaku Kogaku*, vol. 22, p. 780, 1958.

[WIS 87] WISNIAK J., HILLEL J., KATS O., "Holdup and extraction characteristics of jojoba meal", *J.A.O.C.S.*, vol. 64, no. 9, pp. 1352–1354, 1987.

Index

Printed in the United States
By Bookmasters